The Earth Care Annual 1990

The Earth Care Annual 1990

Edited by Russell Wild

 National Wildlife Federation

 Rodale Press, Emmaus, Pennsylvania

The Earth Care Annual **Team:**

Editor: Russell Wild
Research Coordinator: Melissa Meyers
Production Editor: Jane Sherman
Book Designer: Glen Burris
Reprint Permissions Coordinator: Sandra Lloyd
Office Staff: Eve Buchay, Karen Earl-Braymer, Roberta Mulliner
Executive Editor, Health: Carol Keough
Vice President and Editor in Chief, Rodale Books: William Gottlieb

Copyright © 1990 by Rodale Press, Inc.

All rights reserved. No part of this publication may be reproduced or transmitted in any form or by any means, electronic or mechanical, including photocopy, recording, or any other information storage and retrieval system, without the written permission of the publisher.

Printed in the United States of America on acid-free, recycled paper.

If you have any questions or comments concerning this book, please write:

Rodale Press
Book Reader Service
33 East Minor Street
Emmaus, PA 18098

ISBN 0–87857–875–7   hardcover

Distributed in the book trade by St. Martin's Press

2 4 6 8 10 9 7 5 3 1   hardcover

CONTENTS

Foreword ix
Introduction xi

Earth Care for a New Decade

ACID RAIN
3
- Acid Murder No Longer a Mystery—*Audubon* 3
- Smokestacks That Belch—*Omni* 6
- To Treat the Attack of Acid Rain, Add Limestone to Water and Wait—*New York Times* 10
- "River Rolaids" Art Fights Pollution—*Associated Press* 11

AIR POLLUTION
13
- Air Pollution Emerges as a World Problem—*Washington Post* 13
- America's Smog Crisis—*National Wildlife* 14
- A Drastic Plan to Banish Smog—*Time* 16
- Northeast to Try Cutting Fuel Volatility to Reduce Smog—*Christian Science Monitor* 17
- Rapeseed Oil to Fuel German Tractors—*Stockman Grass Farmer* 18
- Huck Finn Meets the Solar Car—*Washington Post* 19
- Can Bicycles Save the World?—*Omni* 21

GREENHOUSE EFFECT
23
- If You Can't Stand the Heat...—*Public Citizen* 23
- Greenhouse Gases Other Than Carbon Dioxide—*Public Citizen* 25
- Ten Steps to Turn Off the Heat—*International Wildlife* 28
- Solutions for an Overheating World—*Environmental Action* 31
- Reactors Redux—*Sierra* 37
- City of Freeways Plants an Idea to Help Nature—*New York Times* 42

OZONE 44

- Saving Our Skins — *Earth Island Journal* 44
- An 11-Point Plan to Save the Ozone Layer — *Earth Island Journal* 46
- How Orange Rinds Help Save the Ozone — *Christian Science Monitor* 49
- Stepping Up to the Ozone Challenge — *Christian Science Monitor* 50
- Vermont Moves to Save the Ozone Layer — *Philadelphia Inquirer* 52

PESTICIDES 54

- A Mother's Crusade — *Organic Gardening* 54
- Start Your Own Chapter — *Organic Gardening* 58
- Big Firms Get High on Organic Farming — *Wall Street Journal* 61
- What's in a Name? — *Wall Street Journal* 62
- Indonesia Cuts Pesticide Use — *Greenpeace* 64
- What Really Happens When You Cut Chemicals? — *New Farm* 66
- Hometown Hero — *Organic Gardening* 69
- Myth vs. Reality — *New Farm* 70
- Hormones to Be Used in Lieu of Conventional Pesticides — *Wall Street Journal* 72
- Please Don't Eat the Daisies — *Newsweek* 73
- Lawn-Care Companies Now Offer Alternatives — *Changing Times* 74

PLASTICS 76

- Persistent Peril — *Organic Gardening* 76
- Pollution Solutions — *Organic Gardening* 78
- Plastic and You — *Organic Gardening* 80
- American Myopia: Focusing on Short-Term Profits — *Omni* 83
- Most Throwaway Plastics Face a Ban in Minneapolis — *New York Times* 85
- A Second Life for Plastic Cups? Science Turns Them into Lumber — *New York Times* 87
- Diaper Wars! — *Environmental Action* 89
- Comparing Bottom Lines — *Environmental Action* 90
- Diaper Solutions — *Environmental Action* 93

SOLID WASTE 95

- Getting into a Heap of Trouble — *National Wildlife* 95
- Sorting Out the Truth — *National Wildlife* 98
- Legislative Avalanche: 2,000 Solid Waste Bills — *BioCycle* 100
- One-Year Moratorium for New Incinerators in Massachusetts — *BioCycle* 101
- Give Us Your Tires, Your Coors Cans... — *Sierra* 102
- The Consummate Recycler — *Environmental Action* 106

CONTENTS

	From Landfill to Landscape—*Christian Science Monitor* 110
	Landfill Alternatives—*Christian Science Monitor* 111
	A Pay-by-Bag Town Thinks Trash—*Philadelphia Inquirer* 113
	Corporations Learn to Turn Their Trash into Cash—*Philadelphia Inquirer* 115
	Recycling Pioneers Are Seeking Ways to Exchange New Life for Old Tires—*Associated Press* 117
	Ontario Recycling in Blue Boxes—*Christian Science Monitor* 118
	Once a Month, an Unruly City Scrapes Itself Clean—*New York Times* 120
TOXIC WASTE **123**	Preserving Our Environment: Preventing Hazardous Waste—*Science for the People* 123
	The Children's Cleanup Crusade—*Sierra* 131
	The Gift That Keeps on Giving...Toxics—*Earth Island Journal* 136
	Freiburg, the Environmental Capital of West Germany—*Earth Island Journal* 138
	Africa Gets Tough on Toxic Dumping—*Christian Science Monitor* 142
TROPICAL FORESTS **143**	Fire at the Equator—*Organic Gardening* 143
	Lovely Cheetahs, Meter-Saved—*Sierra* 150
	For Africa: A Fast-Growing Forest—*Christian Science Monitor* 151
WILDERNESS **153**	With Fate of the Forests at Stake, Power Saws and Arguments Echo—*New York Times* 153
	Postscript: "It's Sort of Kill a Tree, Plant a Tree"—*Los Angeles Times* 156
	A Woman Who Guards a Wilderness—*Christian Science Monitor* 157
	Prop 70: How Californians Won $776 Million to Protect Wildlife, the Coast, and Parklands—*Whole Earth Review* 160
WILDLIFE **165**	Showdown at Sea—*Organic Gardening* 165
	A World without Whales?—*Organic Gardening* 168
	The Return of the Giant Clam—*Earthwatch* 171
	Turtle Rescue—*Omni* 175
	Drought and Wetlands Drainage Take a Heavy Toll on Many Species—*National Wildlife* 176
	People Who Make a Difference—*National Wildlife* 177

A Citizen's Guide to Personal Action

ACTION ALERT   189
Watching the Watchdogs: How to Influence Federal Agencies   189
The Federal Agencies: Who's Who   191
Use the News to Protect Your Environment   194
Making Congress Work for Our Environment   198
Understanding the Congressional Process   199
Useful Information   203

DISPOSAL OF HOUSEHOLD HAZARDOUS WASTE   205

DIRECTORY OF ENVIRONMENTAL GROUPS   211

Index   217

FOREWORD

By Jay D. Hair
President and Chief Executive Officer
National Wildlife Federation

Behold a meadow, alive with the myriad colors of spring wildflowers; a cold, clear mountain stream, teeming with trout; or an elk, strong and alert, testing the waters of a crystal-clean, high-country stream on a warm autumn day. These wonders of nature, and so many others, can be instantly recalled as vivid "snapshots" in the mind's eye of anyone lucky enough to have experienced them firsthand. But in life, they are more like a fragile lens, requiring great care. Abuse or neglect can leave them, and the intricate ecosystems to which they belong, shattered. Once lost, they cannot always be "recalled" with the ease of a photograph or a memory.

The 20th anniversary of Earth Day, April 22, 1990, is a wonderful opportunity for us to recall these and other examples of our magnificent natural heritage, and more important, to rededicate ourselves to the conservation ethic. This book will help you do that.

It's no coincidence that this first edition of *The Earth Care Annual* is being released on this important anniversary. Like Earth Day itself, it is both a celebration and a challenge: a celebration of our great successes in the struggle to rescue our endangered environment, and a challenge to continue the arduous work that further rescue will entail. It is about what individuals, groups, and corporations are doing to repair the environmental "scratches" and "cracks" that cloud our portrait of a clean and healthy planet Earth.

The National Wildlife Federation is very serious about its commitment to the environment. We view the 20th anniversary of Earth Day as a chance to set the agenda for action — to move environmental issues from the margins to the mainstream.

Last year, Americans were repulsed by pictures of an environmental crime scene: coastal beaches that were once summertime playgrounds, littered with garbage, raw sewage, and medical waste; oceans that became dumping grounds when landfills overflowed with billions of disposable diapers and billions more disposable plastic razors. More people began to wonder if our quality of life had become disposable.

Consider these snapshots of our troubled environment.

- A National Wildlife Federation study documented over 100,000 violations of the Safe Drinking Water Act in 1987 alone — violations that put nearly 40 million Americans at daily risk from drinking water that doesn't meet federal health standards.

- Another federation study revealed the escalating chemical assault that industry is waging on the environment. In 1987 alone, 550 million pounds of chemicals were discharged into streams and other bodies of water, 2.6 billion pounds were coughed into the air, 1.5 billion pounds were injected into underground wells, 2.3 billion pounds were dumped into landfills, and 870 million pounds were flushed into municipal wastewater-treatment plants.

- In February of 1989, the Natural Resources Defense Council released a report which concluded that preschool-aged children are exposed to hazardous levels of cancer-causing pesticides commonly found in fruits and vegetables.

- Global warming, caused by various emissions and exacerbated by the destruction of the world's tropical forests, is posing a real threat to our continued existence. It could cause frequent and severe droughts, storms, coastal flooding, and erosion. And tropical deforestation is occurring at a rate of over 27 million acres per year.

- We have absolutely no idea how many species of life share this planet. We do know, however, that through habitat destruction, particularly in the tropics, about 100 species per day become extinct.

There is much more. But you get the picture. We are surely and not so slowly destroying the conditions needed for life to thrive on earth.

Now it is time to stop looking and start moving! All of us, travelers on this spaceship called Earth, must focus on solutions.

The National Wildlife Federation (NWF) is proud to do its part. We have proclaimed the 1990s the "Decade of the Environment," and we have set an ambitious agenda.

NWF is working with Congress to establish a National Energy Policy. We're also developing an "Environmental Education for the 21st Century" project, with the not-so-modest goal of realizing an environmentally literate society.

And, neither last nor least, we are committed to the book you hold in your hands. *The Earth Care Annual* is filled with inspiring, solution-oriented articles, from the story of a new solar-powered car to an update on new methods of pesticide-free farming, from the latest in recycling technology to progressive legislation to save the earth's protective layer of ozone. Each year the annual will present new problems and innovative solutions formulated by people just like you. In fact, if this book has the intended effect, you may very well find your own story published in a future edition!

The annual is published in cooperation with Rodale Press, a highly respected publishing company committed to the well-being and care of the earth. We're proud to have them as partners in this important effort. Now we'd like to have you join our partnership as well!

We hope you find, in these wonderful stories and in the 20th anniversary of Earth Day, the inspiration for a new environmental renaissance. Think boldly! Dream big dreams! We have a chance, perhaps the last chance, to save for our children more than just snapshots of a world that "used to be." It is also our duty.

Our vision must be of a world preserved, clean and healthy for generations to come. That's a picture we can't afford to let fade away.

INTRODUCTION
By Robert Rodale
Chairman, Rodale Press

(Rodale Press)

The Year We Listened to the Earth

This year, we started listening more carefully to the earth. Not just our little patch of the planet. But the whole place. Millions of people—perhaps even billions—began to imagine what life might be like on a terminally sick earth.

Yes, Earth Day in 1970 changed our thoughts, too. But in a different way. That remarkable event put the environment on the agenda for discussion. And for action, too.

People by the millions awoke to the reality of pollution in 1970. And those who became active environmentalists started arguing and fighting for a cleaner way of living. A whole collection of local and national laws requiring cleaner ways of working and living was soon placed on the books.

But that was different. Then, our vision of the problem and the work ahead was quite shallow. We saw the environment then as merely the outer skin over an essentially healthy earth organism. And our challenge in 1970 was mainly to find ways to wash that skin. And to stop doing many of the things that kept making it dirty.

In essence, we thought the earth needed a bath, then set out to find ways to wash it up.

Today, we know that a bath isn't enough. More than just the skin is dirty. Some of the dirt we make is penetrating to the very heart of the earth, and making it sick in very troubling ways. Like changing the climate for the worse. Like making the oceans rise, flooding low-lying cities and even whole countries. And like polluting the air, the soil, and the oceans so much that they lose their capacity to regenerate. Which then means that they could stop nurturing our well-being, as they have done since the dawn of Creation.

Now, 20 years later, we have a far deeper and truly more global vision of the environmental challenge facing us and this earth. Our environmental education is far more complete.

What helped us learn so much about our Earth in just 20 years? What set us on the course to the insight that our ways of living can make or break this whole planet?

Earth Day was Act One. But the overture came the year before, when people first traveled far enough away from this Earth to look back and see the real nature of their planetary home.

On July 20, 1969, the Saturn V rocket that placed Americans on the moon blasted off.

That rocket left an Earth most people still thought of as a platform—a way-station for humanity headed for bigger and better places.

Sure, we needed to keep our way-station clean. But our true future lay somewhere "out there," among the stars.

The first commentators uttered thoughts and feelings that almost everyone then shared.

"At long last, man is on the brink of mastering the Universe," said Walter Cronkite.

And when the eight days of the Apollo 11 mission were over, President Nixon called it "the greatest week in the history of the world since the Creation."

President Nixon came closest to the truth. Yes, it was a monumentally great week. But for a reason no one suspected when the rocket left Florida.

The moon, it turned out, was quite empty of riches. The rocks brought back satisfied our curiosity but did little to increase our real wealth. (I happen to think the rest of space will also turn out to be quite devoid of useful resources, but that is getting too far off the track of this book.)

Something of great value did come back from the Apollo mission, though. An insight of ourselves and our world that was totally unexpected. The cameras and the eyes of the astronauts gave us the first-ever photographs of the entire Earth, showing it to be a living planet of fantastic beauty and health.

There, floating in that sea of emptiness, the photos from space showed this wondrous ball of blue and green, rich in life. Never before had we been able to stand back in space and take a good look at our home. But now we could, and that view made the experience of Earth Day 1970 possible.

That vision of Earth from space made us realize in a new way how lucky we are to be alive here. We also gained a new and deeper sense that our little planet is fragile. Floating in that sea of emptiness, it has but a thin layer of protection—an airy layer of atmosphere—from the sea of deadness that is space.

Something else happened, too. Not only did the new vision-from-space of the Earth let us see its aliveness in totality. Our minds were also prepared for the possibility that this wondrous, special home of ours could get sick all over—just as people and other life forms sicken in a total way.

Yes, surface pollution stayed the focal point of environmental concerns. But we now could see it covering continents—even becoming global in extent. Not just pollution, but destructive changes of a new and unexpected kind. Like the warming of the earth's atmosphere. And the resulting melting of the polar ice caps, releasing vast quantities of fresh water into the seas and flooding low-lying areas.

These events changed our way of seeing the environment around us—and instilled in many new ethics of behavior and responsiveness.

Learning the New Craft of Earth Care

In 1989, a consensus formed that new ways of dealing with environmental problems must be discovered and used.

The idea of 20 years ago that our task is to stop the flow of pollutants and then wash up the remaining mess no longer holds. People are now thinking much more radically. The word radical means to get to the root. We are looking inside the earth for the roots, trying to see and nurture the roots that support us in regenerative ways.

One good example of that trend is the newly popular concept of sustainable agriculture. That means to expand the goal of agriculture beyond just producing food and fiber at a reasonable cost. To do all those things and

INTRODUCTION

more—to also farm regeneratively, making the land and the rural environment better with each passing season.

Sustainable agriculture is a way of making sure that our grandchildren and great-grandchildren—and even their descendants—will have adequate food and fiber. But it is more. Farming is today one of the major causes of pollution of drinking water. And there is great concern about the damage being done to air, land, and wildlife by many of the potent chemicals used so commonly in conventional agriculture. Sustainable agriculture methods are aimed at solving those environmental problems, and by so doing, making sure that our productive land stays productive permanently.

On the American land, times and ideas are indeed changing. Because of this new sense of the value of low-input and sustainable methods, we are in the process of turning a major environmental problem into a source of strength and even regeneration.

Also new to the environment agenda in recent years is the concept of sustainable development. That idea asks an even broader question: How can *all* of our ways of producing last indefinitely? How can they regenerate the renewable resources that the world's growing population depends on so completely?

Aid organizations are concentrating on sustainable development in their Third World improvement work. Even the World Bank—a few years ago almost immune to environmental concern—asks about sustainability. And so do many of the government and private aid organizations. This is a most encouraging development for the environmental community, because ideas and questions about sustainability are environmental tools, just like vacuum cleaners for oil slicks and smokestack scrubbers.

I would go further and say that such ideas are much more powerful tools; they open up a vast new and fruitful territory for earth care and general environmental improvement.

How do these new environmental tools work? What must we do to follow their conceptual meaning into the field and put them to use?

Regeneration is my favorite answer to the sustainability question. By regeneration, I mean using renewable resources in ways that make them better. For example, not just using up the riches of the land growing crops, then waiting some years for the processes of nature to improve the land, but finding new ways to farm so that the crop plants themselves cause the land to get better.

That sounds difficult. Almost like magic. But this earth is in fact still an immensely rich and beneficent place.

Think for a moment of air not just as something that can be polluted but also as a resource of great usefulness.

For example, air contains 78 percent nitrogen. All soils need continual supplements of nitrogen to produce the protein-rich foods we need. Unfortunately, many of the important crop plants can't take nitrogen directly from the air and convert it into food. But leguminous plants can perform that trick, with the aid of special kinds of bacteria that inhabit their roots.

Farmers using these sustainable methods switch away from reliance on chemical forms of nitrogen to the regenerative use of the vast nitrogen resources in the air. They help the environment by reducing our need to burn gas and oil to synthesize nitrogen in factories. Pollution is reduced too, because nitrogen is managed more carefully and less leaches down through the soil to pollute groundwater.

Many thousands of farmers are now using such regenerative methods. And since farming is the largest and most important industry on earth, that kind of thinking is vital to the success of the emerging earth care strategies.

I use air nitrogen as an example because it is a perfect illustration of an internal resource. The term "internal resource" refers to the riches of nature that are within ourselves. When I say "within ourselves," I mean within us personally and available to us free in the parts of nature that are ours—in our immediate neighborhood of earth and sky.

Some people also use other words to say the same thing. "Free ecosystem services" is the term used by Jessica Tuchman Mathews to describe these precious and non-polluting resources in her article "Redefining Security" in the Spring 1989 issue of *Foreign Affairs.* Her view is that environment is now an international security issue, and to achieve the improvement required, we must learn how to use free ecosystem services more effectively.

Here again we get to the key to what has happened to our environmental consciousness at the end of the 1980s. I believe we have gone inside both the earth itself and our own thoughts to find a whole additional agenda for environmental improvement. Think back for a moment to Earth Day 1970. Either then or shortly thereafter, people were going around quoting Pogo's statement that "We have met the enemy, and he is us."

Those nine words were accepted widely as a clear invitation—even an order—to take the environmental challenge to heart as a personal mission. And especially to think about our own personal role as contributors to the general environmental problem.

But what were our options then? Given what we knew 20 years ago about our role as occupiers and stewards of the earth, what could we think of doing?

I have a written record of how I thought. It is a short column telling people what I thought the first Earth Day meant and making some suggestions about what we should do. It was titled "Pollution Is a State of Mind."

Because I felt so good about writing that back in 1970, I took a look at it recently to see if it was worth using again now. Was I disappointed! It was totally out of date. There was an accusatory tone to the piece, very much in tune with Pogo's comic-strip indictment.

Soon, though, I got my disappointment under control. We have made so much progress since then. Today, it is clear to me that the same regenerative tendencies that are in the earth—which farmers are learning to use to create sustainability—are within us, too.

We are not just the enemy. We are also the visionaries. The believers in an earth that can be gardened as well as abused. And the workers that can turn those visions and beliefs into a gradual regeneration of the earth itself.

If Pogo were still commenting on the environment today, I think he would be justified in making a more optimistic pronouncement. Perhaps someday, he could even say, "We have found the solution, and it is within ourselves."

Our New Agenda for Environmental Action

As you read this first edition of *The Earth Care Annual,* try to keep your own regenerative potential in mind. Always realize that as the earth changes—whether for the better or for the worse—you change, too. And the way you perceive the earth changes.

Take inventory, in a sense, of what is within yourself. Ask yourself over and over what strengths and capacity and competence you have to care for the environment around you, and to make it better.

Those of you who have made gardens or planted trees have already been practicing earth

INTRODUCTION

care in the finest and also the most traditional way. But you'll see in this book other good examples as well.

Ten years from now, what will I or someone else be writing to introduce *The Earth Care Annual* for the first year of the coming millennium? Will it be full of good news? Or will the promising insights and regenerative methods that are within our grasp today have failed to reach deeply enough into every person's way of thinking and living?

I am optimistic. True, we have a massive job to do. Many errors of the past need correcting. And there are still vast areas of contamination that must be cleaned and restored. But we now know the importance of doing that work.

The majority of people say clearly that they see a clean, restored environment as perhaps the most important of all our national priorities.

But the main reason for my optimism is the change that is happening within people. We are starting to enjoy the many activities that regenerate us and the environment at the same time. More people are walking for pleasure. And riding bicycles. And not only experiencing the joy of doing those things, but seeing as well how they make a contribution to a better world.

There is so much more to do, but we now at least know where the right track is and we are starting to move forward on it.

Earth Care
for a New Decade

ACID RAIN

Acid Murder No Longer a Mystery

By Jon R. Luoma

Smokestacks That Belch 6
To Treat the Attack of Acid Rain, Add Limestone to Water and Wait 10
"River Rolaids" Art Fights Pollution 11

It's been nearly a decade since acid rain surged into news media and public consciousness with reports from scientists that the rains and snows—once symbols of purity—could quite literally poison entire freshwater ecosystems. Captured in huge weather systems, acid-forming pollutants could travel hundreds or even thousands of miles from their sources to pollute waters across state or national boundaries.

Now, a near-decade and hundreds of millions of research dollars later, concerns about acid rain have broadened to include threats to wider regions of freshwater lakes and streams, threats to forests, and threats to public health. A clear scientific consensus that acid rain is at least a long-term threat to some aquatic ecosystems has solidified.

In many other ways, very little has changed. The key sources of acid rain remain sulfur dioxide from poorly controlled fossil-fuel-burning power plants and nitrogen oxides from a range of combustion sources, including industrial furnaces and cars. The responsible industries, citing high costs, steadfastly oppose tighter controls, instead calling for more research.

Attempts at federally legislating new controls may have

From Audubon, *the magazine of the National Audubon Society, November 1988. Copyright © 1988 by Jon R. Luoma. Reprinted by permission.*

faltered because the pathways of acidification are initially so subtle that widespread damage is not evident. Chemical compounds naturally present in even some of the most sensitive lakes, streams, and watersheds can neutralize acids, often for many years. Only when those neutralizers are used up will a lake quite suddenly begin to turn acid. And although lakes with little remaining neutralizing capacity number in the tens of thousands in North America, lakes actually acidified to date number only in the hundreds. Similarly, research scientists have learned that visible symptoms of forest destruction become obvious only after damage is well under way.

What follows is a compendium of new developments on the acid rain front.

News from Mount Mitchell

When Audubon last visited Mount Mitchell in North Carolina in 1986, plant pathologist Robert I. Bruck was speculating that it might be many years before he had enough data to make clear projections about whether acid rain—or any kind of air pollution—was causing the forest devastation there, so complex was the issue. He said that he'd let us know when he felt scientifically confident to make a bold statement—maybe a decade hence.

"Time's up," Bruck fairly growled in a telephone interview from his prefab headquarters on the 6,684-foot mountain this summer [1988]. "You would not believe how this place has changed even in the time since you were up here."

As it turns out, Bruck and a team of scientists, with their networks of towers, collectors, monitors, tubes, and wires laced through the skeletal trees, have discovered persistent high levels of pollutants—notably ozone and atmospheric acids—that appear to correlate strongly to a forest die-off that has increased by some 30 percent since we visited the mountain.

"It's plain," he says, "that no one has proved, or ever will, that air pollution is killing the trees up here. But far more quickly than we ever expected, we've ended up with a highly correlated bunch of data—high levels of air pollution correlated to a decline we're watching in progress."

A trip up Mount Mitchell, eastern North America's highest peak, is a thumbnail experience of the kind of biotic transition one might see on a 1,500-mile automobile journey northward from the sultry southeastern United States to frigid Labrador.

Although Mount Mitchell sits at about the same latitude as California's Mojave Desert, the trees on and around the blustery mountaintop are hardy near-Arctic species, red spruce and Fraser fir.

And like many of the ridgetop trees along the entire stretch of the Appalachians—like trees in many of the forests of Europe—they are in evident decline. During my 1986 visit, the fir forest at the summit was eerily dominated by death—stark, brown hulks and deadfalls creaking in the ever-present wind.

According to Bruck, it has gotten worse. The damage extends even farther down the mountain into more of the pure red spruce stands, and in some cases, all the way down to the line where hardwood forests begin. According to other reports, Appalachian spruce-fir forests from the White Mountains in New Hampshire to the Great Smokies south of Mount Mitchell are showing signs of reduced growth and general decline.

The information gathered during the federally funded megastudy that Bruck has been coordinating on Mount Mitchell has him saying that he is "90 percent certain" that air pollutants are killing the southern Appalachian ridgetop spruce and fir forests.

Two years ago, Bruck was wondering if some other factor or complex of factors—insects, fungi, climate, or even forest management practices — might be responsible.

But his data now show that more than half of the time, ozone levels on the mountain exceed those at which tree damage has been proven to occur in controlled laboratory studies. Frequently, levels increase to more than double the minimum damage limit. Acidity in the clouds that bathe the summit of the mountain eight out of every ten days has also been extraordinary: ranging from a worst case of pH 2.12 to a best case of 2.9. In other words, on the best days of cloud cover, acidity has been somewhat more than that of vinegar.

"Let me tell you about an experiment 30 yards from where I'm sitting on top of the mountain," Bruck said. He described a set of large outdoor chambers, sealed in clear plastic to create a controlled greenhouse, in which young trees have been placed. The usual, and apparently polluted, mountain air is pumped into one chamber without alteration. A second chamber receives mountain air filtered through activated carbon, which removes ozone and some other pollutants. "We set up that experiment only six weeks ago. We're already getting 50 percent growth suppression in the chamber receiving ambient [unfiltered] air."

He also reports that the research team noted widespread burning of new needle tips on conifers after particularly acid air masses passed through. Analysis of the burnt needles in Environmental Protection Agency laboratories revealed extremely high levels of sulfate, a compound associated with acid rain.

Mountaintops are subject to greater pollutant deposits because they are frequently bathed in polluted cloud water. Extensive forest destruction in Europe began in the 1970s on the mountaintops but has extended to stands at lower altitudes. Many scientists studying the problem feel that air pollution does not kill trees directly but rather weakens them to the point where, like punch-drunk fighters, they are no longer able to withstand normal episodes of moderate drought or insects or diseases that they could otherwise easily resist.

Killer Mosses, Acid Soils

Now come reports from field researchers of problems stemming from acid mosses and acid soils.

Lee Klinger, working with the National Center for Atmospheric Research in Boulder, Colorado, claims to have correlated forest diebacks with the presence of three kinds of mosses: sphagnum, polystrichum, and aulocomnium. He says that the mosses alone are not killing trees but were virtually always present in more than a hundred dying forests that he examined and that they appear to be a key part of a forest death syndrome.

The mosses produce organic acids, which appear to gang up with inorganic acids in polluted rain to mobilize aluminum naturally present but harmlessly bound up in most soils. Additional aluminum appears to be falling out of the atmosphere, bound to dust particles. The mobilized aluminum is toxic to the fine feeder roots of most trees. In fact, says Klinger, mats of killer mosses inevitably overlay networks of dead feeder roots. Furthermore, sphagnum acts as a sponge, saturating the soil just beneath the moss and creating an anaerobic, or oxygen-starved, soil environment, which also helps kill roots. The mosses, he says, occur naturally in forests and may even be part of an extremely slow plant-succession process that, over centuries or millennia, turns old forests to bogs. But the present moss invasion appears to be promoted and greatly speeded up by acidic rainfall.

"In general," Klinger says, "mosses require acid conditions for their establishment — so it appears that as the soils become acidified, there are more places for the mosses to get established." He also suggests that nitrates that form from nitrogen oxide pollutants in the rain fertilize the mosses, which have no roots but are entirely nourished by atmospheric chemicals.

Meanwhile, some scientists have begun to change their minds about acid soils. Until recently, most researchers looking into acid rain's terrestrial effects have assumed that trees and other plants growing in naturally acid soils—including trees found in many northern coniferous forests—had more resistance to pollutants.

But now Daniel Richter of Duke University reports that his laboratory experiments show that naturally acid soils, such as those often found at high elevations in parts of the Appalachians, are highly sensitive to chemical imbalances caused by the addition of more acids from precipitation. According to Richter, highly acid forest soils that become further acidified can become virtually "infertile" through a complex series of chemical reactions that deprives them of nutrients. At the same time, the soils are assaulted by an overload of mobilized and root-toxic aluminum.

It Keeps Getting Worse

New York was one of the first states to become deeply concerned about the effects that acid rain could have on its surface waters, and particularly on the pristine but poorly buffered lakes of the huge and beautiful Adirondack Park. In the 1970s, the New York Department of Environmental Conservation (DEC) jolted the conservation community by reporting that more than 200 lakes, most of them in the western Adirondacks, had already become too acidic for fish to survive, and that many more appeared to be threatened.

Earth Care Action

Smokestacks That Belch
By George Nobbe

Remember when you were a kid and trashed the kitchen table by concocting a secret potion of baking soda and vinegar? If you'd poured the soda down a smokestack instead, you could have done away with a lot of pollution.

Operators of fossil-fuel-burning plants have long been plagued by gaseous emissions, but recent experiments may change that. Researchers have discovered that blowing sodium bicarbonate (baking soda) into smokestacks nullifies many of the discharges, removing between 70 and 90 percent of the sulfur emissions and 10 to 20 percent of the nitrous oxides.

The problem with using baking soda in commercial plants is the cost: Power plants may consume tens of thousands of tons of sodium bicarbonate a year. At $70 to $150 a ton and with high shipping costs, it is an expensive practice. Stuart Dalton of the Electric Power Research Institute in Palo Alto, California, believes companies will be able to afford this sort of cleaning only if a cheaper alternative is found. He points to recent developments in Czechoslovakia, where the National Academy of Sciences has devised a chemical process to create raw soda ash that's just as effective as sodium bicarbonate and far cheaper ($30 a ton). The Czechs are seeking help from U.S. companies to refine their product, and if all goes well, it should be available by 1991.

From Omni, March 1989. Reprinted by permission.

In a new study, three years in the making, the New York DEC reports that fully 25 percent of the lakes and ponds in the Adirondack Mountains are now so acidic that they cannot support fish life. Another 20 percent have lost most of their acid-buffering capacity and therefore appear doomed if acid input continues.

Massachusetts has reported that almost 20 percent—or about 800—of the state's ponds, lakes, and rivers are vulnerable to acid deposition and could become acidified within the next 40 years. Already, according to Environmental Affairs Secretary James Hoyte, surveys have located 217 acidified bodies of water that cannot support natural communities. Particularly alarming to state officials is data showing that more than 50 percent of the state's 34 drinking-water reservoirs have lost much of their acid-buffering capacity since 1940. The largest of these, Quabbin Reservoir, has lost about three-fourths of its buffering capacity, and the Massachusetts Executive Office of Environmental Affairs now estimates that 20 years remain before the reservoir loses all of its capacity to handle acids. Acidification tends to occur within a few years after buffering capacity is lost.

Meanwhile, Environmental Protection Agency researchers, in a recent report, have identified new and surprising acidification sites in the mid-Atlantic states of Virginia, Delaware, Pennsylvania, Maryland, and West Virginia. The report shows that 2.7 percent of sampled stream miles in the mid-Atlantic region are already acidic, with the afflicted number as high as 10 percent at higher elevations. Despite the fact that rain acidity levels in the region are among the highest in the nation, it has long been assumed that soils in much of the mid-Atlantic could buffer the acidity effectively. The study notes that the stream damage is "probably associated with atmospheric acid deposition."

The Pennsylvania Fish Commission estimated in 1987 that half the state's streams will not be able to support fish life by the year 2000 unless acid deposits decline. Now 78 of the 200 members of the Pennsylvania House of Representatives are cosponsoring a bill to slash emissions of sulfur dioxide by more than half, in a program that would be phased in over a period of 13 years.

Pennsylvania has long been in an acid rain quandary. Its emissions of sulfur dioxide are the second worst in the nation, and both its industrial emitters and its powerful coal industry have strongly opposed regulation. But unlike many other high-emission areas, which contain few easily acidified waters, the state has long recognized that many of its woodland streams are acid sensitive. Pennsylvania government insiders expect vigorous opposition from the pollution lobby on the bill: The state coal association issued a statement immediately after the bill was introduced pointing to government "facts" that watershed acidification was not going to get any worse.

Fertilizing Chesapeake Bay

Two-thirds of the excess acidity in precipitation falling on the eastern United States comes from sulfur dioxide. So sulfur dioxide has long occupied the attention of most of those concerned with the problem.

But a new study from the Environmental Defense Fund (EDF) points out that there are plenty of reasons to be concerned about oxides of nitrogen, which not only produce the other one-third of the acid rain problem but are key to the formation of ground-level ozone. (While ozone in the stratosphere is, indeed, necessary to shield the earth from excess ultraviolet radiation, low-level ozone is not only a health hazard but a proven multi-billion-dollar destroyer of agricultural crops and possibly forests.)

The EDF study looked at neither of those

aspects of nitrogen oxides but rather at their ability also to function as fertilizers—in this case, unwanted fertilizers—of the already pollution-ravaged Chesapeake Bay. The resulting report, based on data from government studies, calculated that one-fourth of the total nitrogen entering Chesapeake Bay comes from excess atmospheric nitrogen oxides, which are produced by combustion in automobile engines, in power plant furnaces, and in virtually all high-temperature burners.

The deposited nitrogen, in turn, is one of the key nutrient pollutants feeding algae in the waters of Chesapeake Bay—to such an extent that biological oxygen demand is up and water quality and fish and shellfish survival are down. The entire process is accelerating eutrophication, a rapid and premature aging process in the bay.

According to EDF's analysis, the nitrogen input to the bay from air pollution exceeds the contribution from sewage-treatment-plant effluents and, in fact, exceeds the contribution from all sources but agricultural fertilizer runoff. Although the study was limited to Chesapeake Bay, it suggests that other bays and estuaries may also be suffering from the fertilizing effects of nitrogen raining from the skies.

An Honest Federal Summary?

In the history of acid rain research, surely the strangest few days came in September 1987, when the National Acid Precipitation Assessment Program (NAPAP) finally released a long-awaited "interim report."

NAPAP, set up late in the Carter Administration to coordinate federal acid rain research, had already been criticized by the General Accounting Office for foot-dragging. When the interim assessment was finally released—two years late—it ignited a veritable firestorm of scientific protest.

There were few complaints about the veracity of the three main volumes of the study. But the slim fourth volume, an "executive summary," was at the heart of the heat. J. Lawrence Kulp, then the NAPAP director and a former Weyerhaeuser executive appointed by Ronald Reagan, had written much of the summary himself.

Critics were especially disturbed that the summary's tally of "acid lakes" counted only those so acid that adult fish could not survive, with far less emphasis on the more numerous acidified waters where amphibian and insect life are harmed, fish reproduction is destroyed, and ecosystem food webs are disrupted.

Further, some critics were outraged that the summary promoted as fact an unproven chemical "steady state" theory, favored by Kulp, that lakes in the Northeast would not become more acidic.

Some prominent researchers spoke out. Some of them spoke out vociferously in the pages of *Science,* which had obtained a prerelease copy. Just days after the summary's release, J. Lawrence Kulp resigned.

The new NAPAP director is one James R. Mahoney. Last April [1988], Mahoney offered a pleasant surprise to many conservationist observers. Testifying before a congressional subcommittee, he offered to prepare a new summary. "I believe the executive summary can and should be expanded to be more representative of all the data available," he said, adding that he "would not subscribe...at this time" to Kulp's assertion that acid rain would not harm more northeastern lakes.

Mahoney and Rep. James Scheuer (D.-N.Y.), chairman of the subcommittee, later agreed that a shorter report responding to the scientific criticisms would make more practical sense than a wholly republished "interim" summary. That's because Mahoney has agreed to gear up his staff to accelerate the massive final assessment to meet the original 1990 deadline, despite the previous delays.

NAPAP plans, this time around, to be more active in soliciting criticisms and comments, according to staff ecologist Patricia Irving. "We want this to be a [scientific] consensus document in every sense," she says.

Scheuer, the congressman, appears to agree that the summary, at least, needs some work. He called the 1987 version "intellectually dishonest."

Wait for Clean Coal?

A broad and bipartisan coalition of representatives and senators has introduced compromise legislation in an attempt to break through the political blockade that has stalled all attempts to control acid rain at its source. The new legislation scales back on earlier calls for an annual reduction of 12 million tons of sulfur dioxide. The new target would be 10 million tons per year, or about a 35 percent decrease from current levels.

Connie Mahan of National Audubon Society's government relations office says the society is supporting the bill "even though we are not happy with scaling back by two tons. We went into the 100th Congress believing that the political will for solving this problem was finally taking shape. We continue to hope that by the 101st Congress we'll have acid rain legislation."

The coal and electric power lobbies are continuing to fight legislation which would require them to reduce sulfur dioxide emissions at costs that could exceed $100 million for each poorly controlled fossil-fuel-burning power plant. Instead, they are promoting "clean coal technologies" now in the research and testing stage that could control pollution in the combustion process at much lower costs. However, they have not suggested that large-scale clean coal technology will be available in the foreseeable future.

"We're very suspicious that the promise of future technological improvements is being used as an excuse for not introducing technology that's already known," says Jan Beyea, National Audubon Society senior staff scientist. Beyea points out that the dirtiest power plants, in terms of acid gas emissions, are pre-1980s facilities that were grandfathered at high emission rates by the Clean Air Act. "The real need is for control technology on these older plants. A technology that's useful by the year 2010 isn't going to be of much use. And we don't want to see North America's forests go the way of the European forests."

From his mountaintop headquarters in western North Carolina, Bob Bruck would seem to agree. "People are going to have to start understanding that this is not like some kind of disease where we're going to give all the trees a pill and cure it. We are going to have to decide as a society how to come up with the most logical and reasonable way of implementing what looks like the best solution."

Earth Care Action

To Treat the Attack of Acid Rain, Add Limestone to Water and Wait

By William K. Stevens

In a nascent counterattack on the effects of acid rain, American scientists and wildlife experts are increasingly bringing dead lakes and streams back to life by using the simple technique of adding limestone to the water.

The technique, long used in Sweden, is not a permanent solution to the problem of acidification, experts say. That will come only when the industrial emissions that cause acid rain are sufficiently reduced, even proponents of the treatment method say. But they believe that "liming," as it is called, now offers a practical way to stave off some damage caused by acid rain until there is a permanent solution. It also restores the health of some lakes and streams where life has already been destroyed by acidification.

"It works, but it's a stopgap," says Dr. Patricia Bradt, a biologist at Lehigh University who is an authority on the subject and who has recently completed a study of the effects of liming on lakes in the Poconos.

Expanding Movement in U.S.

Liming has been tried in a few spots, including the Adirondacks, for years. But now the movement appears to be growing. From helicopters and barges and pumps in the warm months, and from fertilizer spreaders on the ice in winter, fisheries experts are applying granulated limestone to waters in New England to the Adirondacks and the Great Lakes states and south into Appalachia. An estimated 100 such projects are now under way, a bare beginning as against the large-scale Swedish program, in which about 500 lakes a year are limed.

Nevertheless, large-scale or small, the effects are reported to be dramatic. The alkaline character of the limestone rapidly offsets a lake's acidity, rapidly restoring the water to a neutral condition, or nearly so. In one experiment in the western Adirondacks, "we added fish four days after liming" and they flourished, according to Dr. Donald Porcella, an environmental scientist with the Electric Power Research Institute, which conducted the experiment.

From the smallest algae to the biggest fish in the food chain, the populations of acidic or acidifying lakes gradually reestablish themselves after liming, although restocking of fish speeds up the process.

Dr. Bradt's study involved two 50-acre lakes, White Deer Lake and Bruce Lake, about ten miles apart in Pike County, Pennsylvania. The two displayed similar characteristics when the study began in 1984. Both lakes had been declining because of acid rain, the investigators found.

One Lake Treated, One Untouched

Bruce Lake was not touched. White Deer Lake was treated with 100 tons of limestone, distributed by a tractor-drawn fertilizer spreader on the lake's frozen surface in February 1985. For about 2½ years, this application held acidification at bay. When it became evident that the lake was acidifying again, 15 more tons of limestone slurry, granules mixed with liquid, were pumped into the lake from a barge.

In those first 2½ years, Dr. Bradt reported, acid-sensitive organisms like mayflies, clams, snails, dragonflies, caddis flies, zooplankton, and diatoms, or microscopic algae, all reestablished themselves. But in the untreated Bruce Lake, such organisms steadily declined. At the end of the 4-year experiment in 1988, Bruce Lake had become acidified, while White Deer Lake had not.

In the Adirondacks, Dr. Porcella and a research team limed three lakes beginning in 1983, two of them fishless and one with lake trout that, because of acid rain, were showing signs of being unable to reproduce. The two fishless lakes were stocked with brook trout,

---

Earth Care Action

"River Rolaids" Art Fights Pollution

Artist Buster Simpson has a cure for America's polluted waters. The cure is large limestone tablets—he calls them "river Rolaids" or "Tums for Mother Nature"—and he's just dumped a bunch of them in the water fountain at the Hirshhorn Museum in Washington, D.C.

In his slightly wacky fashion, Simpson is trying to prove a point.

The 40,000 gallons of water that circulate through the fountain's pumping system come from the Potomac River. His limestone antacid pills slowly neutralize the acidic content of the water, sweetening it enough to nourish delicate elm and sugar maple saplings he has placed in the fountain.

Both types of trees are threatened by acid rain, he notes.

"These large pills have become a stop-gap solution," he says. "The bigger the problem, the bigger the pill."

The pills are nearly two feet in diameter and weigh more than 42 pounds. Before bigger pills are needed, he suggests, "we ought to get to the source and clean up the smokestacks."

Artist Buster Simpson stands in his fountain masterpiece. (AP/Wide World Photos)

Simpson, a Michigan-born artist from Seattle, is using the huge, circular fountain in the Hirshhorn's inner courtyard to create a somewhat bewildering sculptural message about the folly of man's destruction of the environment.

Simpson's creation includes a semicircle of large plastic drums connected to gushing fire hoses, two aluminum canoes filled with potted saplings, a ring of submerged limestone tablets and—next to the fountain—a satellite dish and an aluminum plate bearing the likeness of George Washington.

"I'm playing with metaphors," Simpson says. "This whole piece is about our closed system, our planet, just as the fountain is a closed system. What we spew out, we will take in again. This deals with acid rain, the carbon cycle, ozone, the biosphere."

From the Associated Press, May 11, 1989. Reprinted by permission.

and the fish "have done quite well," said Dr. Porcella, adding that "they have grown fast and are healthy fish." In the third lake, the lake trout resumed reproduction.

Similarly positive results are being reported from other projects in Massachusetts, Rhode Island, Pennsylvania, West Virginia, Michigan, and Minnesota. But for all the successes, there are major limitations as well.

Need for Repeated Limings

First, lakes must be periodically limed again because their water is constantly being replaced, washing out the limestone. Some lakes are considered unsuitable for the liming because they flush too rapidly.

Liming is generally cost-effective in lakes that flush themselves out no more often than every six months, said Dr. Timothy Adams, a chemist who is a consultant for Living Lakes Inc., a Washington-based nonprofit organization that has taken the lead in the renewed effort to promote liming in the United States.

But, he said, costs must be judged on a case-by-case basis, taking into account the importance of the lake to the economy and the environment.

Dr. Bradt believes that liming "should be considered an expensive Band-Aid on a gaping wound."

Still, that Band-Aid is being applied more often. In the last three years, Living Lakes has limed 28 lakes. It is preparing to lime 10 more this year, and has helped support a number of research projects, including Dr. Bradt's. The organization is supported by the utility and coal industries as well as by some environmental groups, according to Dr. Robert Brocksen, a specialist in aquatic ecology who is executive director of Living Lakes.

"We Don't Solve the Problem"

Actual treatment costs run from about $10 to about $15 an acre per year, Dr. Adams said. For instance, he said, the estimated treatment cost for a newly treated 350-acre lake on the Massachusetts/Rhode Island border, called Wallum Lake, is about $4,200 a year. The cost is based on an estimated flushing rate of once every two years.

In Sweden, the government is spending $35 million this year [1989] to lime that country's lakes, and expects to spend $509 million by 1992. "Lake liming is not research any more," said Dr. Harald Sverdrup, a professor of chemical engineering at Sweden's Lund University, who is a consultant to the government on the matter. "We know how to do it, and in terms of aquatic management, it works."

"We don't solve the problem this way," Dr. Sverdrup said in a telephone interview, "but we take away from the worst effects. That buys time for us to work for emission reductions."

A few years ago, liming was criticized in the United States as representing an attempt by the utility and coal industries to avoid having to invest in devices to control acid rain. "There might be some people who still view it as that," said Dr. Brocksen. "If they do, they're incorrect." Liming, he said, is "something that's got to be done no matter what happens, and it's something that can be done now."

From the New York Times, *January 31, 1989. Copyright © 1989 by the New York Times Company. Reprinted by permission.*

AIR POLLUTION

Air Pollution Emerges as a World Problem

By Robin Herman

America's Smog Crisis 14
A Drastic Plan to Banish Smog 16
Northeast to Try Cutting Fuel Volatility to Reduce Smog 17
Rapeseed Oil to Fuel German Tractors 18
Huck Finn Meets the Solar Car 19
Can Bicycles Save the World? 21

The industrialized nations heard the alarm in 1952. That was the year a terrible smog settled over London, making breathing so difficult that 4,000 people died. The lethal pollutants were sulfur dioxide and smoke from the tons of coal burned to heat homes and fuel industry.

Much has been done since to clean up the air over the world's big cities, but a new, detailed report by the World Health Organization (WHO) and the United Nations Environment Programme shows that two-thirds of the world's city dwellers still live in unacceptably polluted environments, and the situation is getting worse.

Milan, Seoul, and Rio de Janeiro are coping with some of the world's highest levels of sulfur dioxide. Dust and smoke hovers thick over Kuwait, New Delhi, and Beijing. Parisians breathe air laden with carbon monoxide, at times more than triple the amount acceptable under WHO guidelines.

On the other end of the scale, Vancouver, Copenhagen, and Melbourne stand out as among the world's cleanest big cities. New York, despite its reputation for grime, falls well within the organization's guidelines for most pollutants and could be ranked in the middle range along with London and Tokyo.

From the Washington Post, *October 11, 1988. Reprinted by permission.*

Through their Global Environment Monitoring Systems (GEMS), the two U.N. organizations have been gathering information on pollution for 10 to 15 years in more than 60 countries. Some large areas of the developing world and some important industrialized countries—such as the Soviet Union—have been left out. Nevertheless, Dr. Michael Gwynne, head of the GEMS system, said the report, covering air, water, and food contamination, represents "the most comprehensive global study ever carried out."

The report, released last month [September 1988], concludes that while the industrialized nations have been able to reduce emissions of pollutants and improve air quality in their cities, that progress is being offset by the rapid urbanization and industrialization of the developing world. The unplanned growth of cities is inevitably accompanied by increased traffic, energy consumption, industrial activity, and pollution.

Population projections tell the tale. In 1950, there were only 13 cities in the world with

America's Smog Crisis

During the long, hot summer of 1988, much of America's air turned thick and rancid with smog—ozone pollution created by the action of sunlight on such emissions as automobile exhaust. The worst air pollution measured in a decade, the smog obscured some heartening statistics on the continuing reduction of lead and carbon monoxide from the nation's environment.

Throughout the year, such traditionally car-choked cities as Los Angeles, Houston, New York, and San Diego continued to have monumental problems. But the smog of 1988 was alarmingly extensive. The Detroit region, for instance, experienced serious ozone pollution above the federal health standard on more days in 1988 than in any year since 1978—nearly ten times as many as in 1987. In Chicago, ozone pollution reached serious levels on twice as many days as in 1987, often soaring to nearly twice the federal standard and triggering the city's first major smog alerts in ten years.

Even worse, the smog appeared in record proportions far from its usual city haunts. For the first time since measurements began, for example, excessive ozone levels were recorded in rural Maine and northern New York State. And, according to a study released last summer [1988] by the World Resources Institute in Washington, D.C., such ground-based ozone is causing some $5 billion in annual U.S. crop losses.

The smog crisis tended to obliterate the progress of recent years in reducing certain automobile emissions. The Environmental Protection Agency (EPA) reported last year, for instance, that the average level of lead in nationwide air samples dropped 35 percent in 1986, the sharpest yearly decline for any pollutant since monitoring began in the early 1970s.

Although new cars run cleaner, the number of automobiles in use is increasing twice as fast as the country's population, and the number of miles driven per year is approaching two trillion, an increase of 20 percent between 1980 and 1987. To make matters worse, late last year the federal government weakened the automobile fuel-efficiency standards, allowing more pollution per mile in the future.

The dimensions of the problem added new urgency to efforts of the federal government to make some long-delayed decisions about the implementation of the Clean Air Act. Amendments to the act passed in 1977 required all U.S. jurisdictions to attain certain air quality standards in ten years. As the deadline approached, with more than 100 cities unable to meet requirements, Congress granted yet another extension, to August 31, 1988.

At that point, the EPA imposed its first penalty—a ban on construction of any facility in the Los Angeles area that would emit more than 100 tons of pollutants a year. The measure has no effect on existing industrial polluters or automobiles. But it did serve notice that communities without at least a credible plan for solving their air problems do face consequences, including loss of federal funds for highways and sewage treatment plants.

From National Wildlife, *February/March 1989. Reprinted by permission.*

populations over four million. By 1980 there were 35 big cities—most of them in the developing world. By the year 2000, urbanization in the Third World will push the total to 66 cities of more than four million, and the world's largest cities will be Mexico City and São Paulo, Brazil. By 2025, there will be 135 such cities.

The study monitored five basic air pollutants—sulfur dioxide, suspended particulate matter (dust and smoke), carbon monoxide, lead, and nitrogen dioxide.

Sulfur dioxide pollution is formed primarily from the combustion of coal, oil, and other fossil fuels. It is a potent irritant and leads to respiratory illness.

Among 54 cities monitored for sulfur dioxide, Milan had the highest annual average levels—185 micrograms per cubic meter of air. The WHO guideline is no more than 40 to 60 micrograms. Shenyang, an industrial city in China, had similar values. Paris registered 83 micrograms, while New York had 52.

"Milan is in a valley, and it gets temperature inversions very easily," explained Ann Willcocks at the U.N.-sponsored Monitoring and Assessment Research Center in London. "That traps pollutants. Los Angeles is also in a basin and has the same problem."

The air used to be much worse over Milan. The city has been working hard to reduce sulfur dioxide. The concentration has dropped by more than 8 percent annually during the 1980s, one of the best improvements among cities surveyed. This trend is typical of industrialized nations that can afford the cost of pollution-control technology.

In suspended particulate matter, Kuwait and several Chinese cities—Beijing, Xian, and Shenyang—have major problems. The WHO guideline for particulate matter is an annual average of no more than 60 to 90 micrograms per cubic meter of air in order to protect against chronic health effects related to long-term exposure. In conjunction with sulfur dioxide exposure, high particulate exposure can lead to respiratory problems.

In Kuwait, the concentration was measured at 603 micrograms, and in Beijing, 399. The count in Chicago was 96. New York, with a count of 57, fell within the guidelines.

However, the report emphasized that most of the cities with the highest levels of particulates are in geographical areas where natural wind-blown dusts make a major contribution to the count. Dust is not nearly so dangerous as particles under ten micrometers in diameter, which are small enough to be inhaled into the lungs. This kind of small particle is caused primarily by combustion.

China's pollution problem is due to the use of coal for home heating and industry, accounting for over 70 percent of energy production in that country.

"Much of this coal is not very good coal," said Guntis Ozolins, manager of the WHO's Prevention of Environmental Pollution unit. "It is high in ash and high in sulfur, and for that, you need some control—to scrub it from the stack or process the coal before it's being used. China is just starting to work on this. They are taking things seriously, but it's going to be a major effort."

Carbon monoxide is one of the most widely distributed air pollutants. The gas restricts the binding and transport of oxygen in the blood and can be lethal at accidentally high doses in enclosed spaces. The largest artificial source of carbon monoxide is motor vehicle emissions, also the source of most lead pollution.

The report cautioned that its values for carbon monoxide were not as reliable as those for other pollutants. Concentrations of carbon monoxide drop off sharply once one moves away from the curbside, thus the precise location of the sample site can greatly affect the reported values. Also, variations over short distances within a city can be as great as those between cities.

Given those caveats, Paris was at the top of the scale in a survey of 20 major cities. It had seven times more carbon monoxide in the air than did London. Between 1980 and 1984, readings regularly showed more than three times the WHO guideline for concentrations in an eight-hour period.

Paris also ranked highest among 23 cities in annual lead averages, with 1.9 micrograms per cubic meter of air, above the WHO guideline of 0.5 to 1.0. Lead interferes with the production of meoglobin, the molecule that carries oxygen in the bloodstream, and can be toxic to the nervous system at high levels. France has not removed lead from gasoline and, along with the rest of Europe, has less stringent emission standards on cars than do the United States and Japan.

The lowest national average level of lead was found in Japan, just 0.1 micrograms. Gasoline has been virtually lead-free in Japan since 1976. In the U.S., all 220 urban sites in a national network reported annual average lead concentrations below 1 microgram in 1980.

For nitrogen dioxide pollution, the report found that all cities surveyed fell under the annual average standard set by the U.S. and Canada. Lacking data on the effects of long-term nitrogen dioxide exposure, the WHO has yet to set a guideline value.

Earth Care Action

A Drastic Plan to Banish Smog

By Philip Elmer-DeWitt

The first gray-brown stains appeared in the azure skies above Los Angeles before the outset of World War I. During World War II, the summer haze was beginning to sting the eyes and shroud the mountains that ring the city. By the mid-1950s, Los Angeles' smog, as the noxious vapor had been dubbed, was sufficiently thick and persistent to wilt crops, obstruct breathing, and bring angry housewives into the streets waving placards and wearing gas masks. Oil companies were urged to cut sulfur emissions. Cars were required to use unleaded gas, and exhausts were fitted with catalytic converters. But as the city continued to grow unabated, so did its choking smog.

Now, after more than 30 years of struggling to clean up what has become the nation's number one air pollution problem, California officials have taken decisive action against the primary source of the trouble: the unfettered use of fossil-fuel-burning private vehicles in a city that has long been in love with the automobile. By a vote of ten to two, the directors of the south coast air-quality-management district, a regional agency with authority over Los Angeles, this month [March 1988] adopted a sweeping 20-year antipollution plan. It will not only drastically curtail automobile use in the Los Angeles basin but also convert virtually all vehicles to the use of nonpolluting fuels by 2009. "The public is ready for change," declares Jim Lents, executive officer of the management district. "This plan signals the beginning of that process."

The proposal, referred to simply as the L.A. Plan, is 5,500 pages long and three feet high, and was five years in the making. It calls for elimination of 70 percent of smog-producing emissions in the Los Angeles area by the year 2000. In the plan's first five-year phase, 123 separate regulations will ban the use of aerosol hair sprays and deodorants and require companies, regardless of the cost, to install the best antismog equipment available. But one of the plan's primary objectives is to break the city's addiction to the internal-combustion engine. First, it imposes stricter emission standards and forces employers to encourage car pooling. Then it calls for conversion of most vehicles to methanol and other cleaner-burning fuels. Finally, in a Buck Rogers phase that assumes rapid advances in fuel-cell technology, it calls for a massive switch to cars, buses, and trucks powered by electricity.

"It's quite a remarkable achievement," says David Howekamp of the Environmental Protection Agency (EPA). Adds Richard Ayres, chairman of the National Clean Air Coalition: "It's a bold attempt to grapple with the real pollution problems." The EPA is expected to approve the Los Angeles plan and use it as a blueprint for a federal program that will include cities like Chicago and New York.

From Time, *March 27, 1989. Reported by Gisela Bolte/Washington and Sylvester Monroe/Los Angeles. Copyright © 1989 the Time Inc. Magazine Company. Reprinted by permission.*

Earth Care Action

Northeast to Try Cutting Fuel Volatility to Reduce Smog

By Mark Trumbull

While southern California's plan to improve air quality may be the most wide-ranging in the nation, it is not the only region struggling to meet federal standards.

Seven states in the Northeast are hoping a new regulation on gasoline will be a major step toward compliance with federal clean air regulations.

Five New England states, as well as New Jersey and New York, have separately passed laws to decrease the vapor pressure of gasoline to reduce the amount that evaporates—before the engine burns it—to become smog.

The states say they can attain a 7 percent reduction in the release of smog-creating chemicals by automobile traffic if allowed to lower the vapor pressure of gasoline from around 11.5 pounds per square inch to a 9.0 level during summer months.

If the plans are approved by the Environmental Protection Agency (EPA), the seven states would in effect create a new regional standard to reduce smog levels that sometimes exceed federal limits in summer. Heat from sunlight accelerates the chemical reaction that forms ozone smog. The brown haze is produced as nitrogen oxides (pollutants emitted from car exhaust) mix with evaporative chemicals that escape from a car's fuel system without being burned.

The EPA recently introduced a national standard of its own with a ceiling of 10.5 pounds for the summertime. But the agency has already proposed waiving its regulation in five of the seven states to permit the tougher standards. California has used a standard of 9.0 since 1971, according to William Sessa of the state's Air Resources Board. But EPA administrator William Reilly has not yet given official approval for a 9.0 standard for any of the northeastern states.

Richard Wilson, director of the agency's office of mobile sources, says the issue is not whether the new standard is a good idea, but when refiners will adapt.

"The issue is timing," Wilson says, noting the EPA itself plans to impose a 9.0 standard nationwide in 1992 but believes an earlier transition would not give oil companies ample lead time to install new refining equipment.

Even the regional 9.0 standard in the Northeast would force changes that would take at least 65 days for most oil companies to cope with, says David Deal of the American Petroleum Institute.

The short-term result might be an increase in imported gasoline, which already makes up about one-third of the supply for the region, according to Dennis Eklof, an analyst with Cambridge Energy Research Associates. Still,

---

Earth Care Action

Rapeseed Oil to Fuel German Tractors

Scientists in Braunschweig, West Germany, have perfected a method of "transesterifying" rapeseed oil so that it can be used as fuel in diesel engines with no buildup of heavy deposits that have limited the use of vegetable oils in diesel engines up to now.

Environmentalists are excited by the process because it allows a significant decrease in stack soot and a sulfur emission of nearly zero. Also, although engines fueled by rapeseed oil do emit carbon dioxide, this will be offset by new crops of rape plants, which, like all growing plants, consume carbon dioxide. A balance is struck, and the amount of carbon dioxide in the air remains constant.

Experimental tractors at the Braunschweig research establishment have been run for thousands of hours on pure rapeseed ester oil without any disturbances at all.

The initial use for the rapeseed fuel oil is as a replacement for petroleum in West Germany's many farm tractors. The government estimates that it would require some 15 percent of the country's farmland to grow the rapeseed needed to fuel the tractors.

The German scientists point out that this acreage is virtually the equivalent of the acreage that was needed for draft animal feedstuffs prior to the diesel tractor era.

Because a high-quality, high-value glycerin is produced as a by-product of the transesterification process, the conversion costs from raw vegetable oil to fuel oil are insignificant, the scientists say.

From the Stockman Grass Farmer, *March 1989. Reprinted by permission.*

Eklof does not agree with Petroleum Institute forecasts that the switch would severely strain the region's refining capacities.

If the new regulations go into effect, gasoline prices in the Northeast are likely to rise. Official estimates put the increase at about 2¢ per gallon. But officials see this as the least expensive first step to improve air quality.

One area where Massachusetts expects further significant reductions in pollutants is at the gas pump, where it hopes to mandate rubber "boots" around pump nozzles that recover vapors that would otherwise escape. California already uses such a system.

Reprinted by permission from the Christian Science Monitor, *April 14, 1989. Copyright © 1989 the Christian Science Publishing Company. All rights reserved.*

Earth Care Action

Huck Finn Meets the Solar Car

By Ellen Goodman

The Tour de Sol isn't as famous as the Indianapolis 500. Not yet anyway. Nor is James Worden a race driver of the same stripe as Emerson Fittipaldi. Indeed, his car went more than 100 miles an hour slower than the Penske PC-18's average of 167 miles per hour that won the famous prize. And his pit stops are a bit longer as well. Worden has to wait for the sun to charge up the batteries.

Nevertheless, on Memorial Day weekend [1989], while millions of Americans watched 500 ear-splitting, gas-guzzling, air-polluting miles end with a bang-up finish, James Worden slipped over the line known as Nerd Crossing at MIT. The 22-year-old had won the 210-mile, three-stage race from Montpelier, Vermont.

The Cambridge crowd on hand to greet him was small by Indy standards, but the victory was pure, sunny, and sweet. Worden isn't just going for a trophy, after all. When he is graduated from MIT, this young scientist isn't going into supercomputers or the military-industrial complex.

He and his teammates are going to pursue what he unabashedly calls the dream: "We want to mass-produce solar cars." He doesn't blush at the suggestion he might become the next Henry Ford. It's his straight-ahead plan.

So to those who have become dubious about the trade-offs of science, cynical about progress, sure that there is a price to be paid for every advancement, the James Wordens of the world are as refreshing as a commencement day. He is one of the breed of high-tech environmentalists, people who don't think we have to choose between modern life and the ozone layer. Who believe that we "can do," without screwing up.

Not surprisingly, Worden made his first solar car while he was in high school. It is memorialized, or at least stored, in the garage behind the house in Arlington, Massachusetts, where he grew up, the middle of five children born to a lawyer and a microbiologist. On the day after victory, the Worden home bore a flag in the front and a nest of finches in the back. But in the driveway was the Worden center-

A contestant in the Tour de Sol "races" across the Connecticut River. (Chris Gee/*Times Argus*)

piece: a solar-powered commuter car. It's a commuter car, not a racing car, that he wants to produce.

"It is kind of crude, and it's really dirty," Worden apologized as he ushered me in for a test drive through the neighborhood. This was a five-year-old version, aluminum on the outside, radio and tape deck on the inside.

The car runs on silicon solar cells, a battery, and a motor. One full day of sun will store up to 50 miles of power. Not a lot, but then the average commute to work is 10 miles. This crude model goes 35 miles an hour, a fact that Worden proves as we tool the neighborhood.

But the real beauty, he says, is that "if you're sitting in a traffic jam in the sun, this car isn't using power, it's gaining power. There's no pollution, no noise. That's the most amazing part."

There is something about this scene that sounds too Huck Finn-ish, too backyard-ish, to compete in the modern world of GMs and Fords. But Huck Finn didn't go to MIT, and Worden's team is not the only one that has seen the handwriting on the wall or the hole in the ozone.

"The automakers are saying they can't meet the lower emission standards," Worden says. "The environmentalists are saying that low emission standards aren't good enough. I put A plus B together and say we have to have solar cars."

So this summer, along with Anita Rajan, his classmate, senior project adviser, and fiancée, Worden's Solectron Corp. will begin to build a "sleek-looking," two-seat sports car that they hope will have a range of 200 miles and a speed of 60 miles an hour. In its early stage, they expect it to sell for $15,000.

The idea of solar cars has been around. So have the prototypes, including GM's $8 million version. But the hot cars on the market this year are as archaic as fossil fuels: big and bad. They are in lethal conflict with the environmental laws popping up in places as distant as California and Vermont.

The conventional car is the environmental equivalent of the cigarette. It's the equal-opportunity pollutant. Every one of us gets a chance to pollute the air on the way to the grocery store. The Big Three car makers may be too big and too few to change. Perhaps it takes environmental entrepreneurs in their backyards to start tinkering with solutions.

But before he begins manufacturing, James Worden is getting ready for one more race. "If I can cross the country in ten days on the sun, then, by golly, anybody can go to work on the sun." When you get there, simply pick a sunny place to park.

From the Washington Post, *June 6, 1989. Copyright © 1989 the Boston Globe Newspaper Company/Washington Post Writers Group. Reprinted by permission.*

Earth Care Action

Can Bicycles Save the World?

By Jane Bosveld

"Great Britain isn't as advanced as we are," wrote an American student in 1967. "Probably half the people still ride bicycles." The student was wrong about bicycle ridership in Britain— only about one in four Brits owns a bike, and most ride for leisure, not as alternative transport. But the American student's observation reflects the prevailing attitude in industrial nations that bicycles are somehow second-class vehicles, dwarfed by the power and convenience of automobiles. A bike, of course, won't win any contests of speed or long-distance commuting, but in an age when the implications of pollution threaten the health of the world, human power is looking better and better.

Bicycle lovers have longed for the day when their machines would regain the respectability lost after horseless carriages took over the streets. They have proselytized about the joys of bike riding: the closeness one feels to nature when pedaling through the countryside; the exhilaration one feels at making it up a long hill or skillfully maneuvering through a busy intersection. Biking, they contend, is the one exercise suitable for just about everyone. When you ride a bike, the bike bears the weight of your body, allowing you to exercise your muscles without taxing your joints. Many an individual suffering from arthritis or knee trouble has turned to the bicycle for relief.

Bicycle activists have mobilized in most North American and European cities, lobbying transportation departments for bike lanes and trying to rustle up support from nonbicycle riders. Despite these efforts, however, transportation planners have remained notoriously unsympathetic to the needs of bikers. It is a stance we may all come to regret. Consider these facts published in a recent article by the Worldwatch Institute, a major think tank for environmental conservation.

- Gasoline and diesel fuel emissions are major contributors to acid rain and the depletion of the ozone layer. They are also linked to about 30,000 deaths each year in the United States alone. Interestingly, the worst pollution comes from short car trips, because a cold engine is particularly inefficient, releasing a high percentage of unburned hydrocarbons into the atmosphere. Many of these short trips could easily be done on a bike.

- If just 10 percent of the Americans who commute to work by car rode their bikes to work or to a train or bus that would take them to work, more than $1.3 billion could be cut from the U.S. oil import bill. (Oil imports account for nearly a quarter of the country's $171 billion trade deficit.)

- Most cities devote at least one-third of their land to parking lots and roads. In the United States this comprises more land than the entire state of Georgia.

"In their enthusiasm for engine power," writes Marcia D. Lowe, author of the Worldwatch article, "transit planners have overlooked the value of human power. With congestion, pollution, and debt threatening both the industrial and developing worlds, the vehicle of the future clearly rides on two wheels...."

Although there are more than twice as many bicycles (800 million) as cars in the world, most of them are used for transportation only in countries like China and India. A few industrial nations have embraced the bicycle as a workhorse: The Netherlands, for example, has more than 9,000 miles of bicycle paths, and one city in Japan has actually built a 12-story bicycle parking lot, using cranes to lift and park up to 1,500 bikes at a time. In the United States, however, bicycles are viewed essentially as pleasure machines, to be dusted off during the summer months for Sunday rides in the park.

But what would happen if we began to use bicycles more frequently? If, say, we hopped on a bike to go get a gallon of milk or to visit friends on the other side of town? What if we saved the car for big hauls and long trips? Before this can happen, of course, much must be done to make bicycle riding safe and pleasurable. Biking may be wonderful exercise and environmentally sound, but few individuals will be willing to pedal down roads where cars and trucks zoom past them with inches to spare, leaving the biker to wobble in a blast of air. Until roads are built with bike lanes or at least wide shoulders, few people are likely to get in the habit of biking. Even with those improvements, it will take a shifting of attitudes to get most people to take up two-wheel travel. People will need to believe that even one less trip in the car adds up to something, that riding a bicycle is, like recycling paper or conserving electricity, an endeavor worth pursuing. Deciding to ride a bike is taking on responsibility. Not everyone will choose to do so, but for everyone who does, the world, rest assured, will be at least a little better off.

From Omni, *February 3, 1989. Reprinted by permission.*

GREENHOUSE EFFECT

If You Can't Stand the Heat...

By Annie Eberhart

Greenhouse Gases Other Than Carbon Dioxide 25
Ten Steps to Turn Off the Heat 28
Solutions for an Overheating World 31
Reactors Redux 37
City of Freeways Plants an Idea to Help Nature 42

The heat and drought that other regions of the globe experienced during the past few years hit the U.S. with a vengeance this summer [1988], halting transportation on the Mississippi, biting deeply into our grain reserves, and ravaging our western forests with some of the worst fires in memory.

This decade has been unusually hot—the four hottest years on record were during the 1980s—and the blazing heat of this last summer has focused attention on the continuing debate among professional weather watchers over whether we are experiencing a permanent rise in the global temperature or just an awful aberration. There is no conclusive evidence that the heat wave of 1988 is part of a long-term warming trend. There is, however, mounting data to suggest that human beings are changing the composition of the atmosphere with pollutants and destruction of the earth's vegetation. The result is difficult to predict with certainty, but scientists speculate that it could lead to dramatic changes in the world's climate.

In a natural phenomenon known as the greenhouse effect, gases and water vapor literally form a blanket around the earth, trapping infrared heat reflected from the planet that would otherwise wander off into space. Without the greenhouse effect,

From Public Citizen, *November/December, 1988. Reprinted by permission.*

the world would be 33°C colder and covered with ice, according to Daniel Albritton, director of the Aeronomy Laboratory of the National Oceanographic and Atmospheric Administration, the government's weather forecaster.

But now that blanket of gases is getting thicker, and this is why scientists theorize that the earth's temperature is rising. "We have begun to perturb the planet's [natural] greenhouse effect," said Albritton in testimony before a congressional committee last summer. "Namely, human activities are, without a doubt, changing the abundance of some of the atmospheric trace gases that are responsible for the trapping of infrared radiation. As a result, we are altering the heat balance of the planet."

Scientists estimate that since the beginning of the industrial era, the earth's average temperature has increased between 0.5° and 1.5°C, and could go up 4.5° by the year 2030. "Not since the dawn of civilization some 8,000 years ago has Earth been about 1 degree warmer than today," says the World Resources Institute in its 1987 report *A Matter of Degrees: The Potential for Controlling the Greenhouse Effect*. "To find conditions like those projected for the middle of the next century, we must go back millions of years."

An enhanced greenhouse effect could lead not only to a hotter earth but also to the possibility of havoc created by abrupt climate changes: tumultuous storms, the extinction of animal and plant species that cannot make the adjustment, the loss of agricultural lands to desert, and the flooding of coastal regions. What's worse, according to the experts, the global warming that has begun cannot be reversed. The best we can hope for is to diminish the trend.

The prospect of global warming is a crisis so immense that it would seem as though the individual is incapable of doing anything about it. But in fact there are a number of concrete steps individuals can take. Public pressure on local, regional, national, and international policymakers is necessary to force an immediate reduction in the amount of gases we introduce into the atmosphere. Further, there are a number of changes individuals can make to reduce their own contributions to global warming. Advocacy and actions by individuals can provide a model for others to follow, exponentially increasing the number of citizens pursuing solutions to the ways we pollute our atmosphere.

Cheap Oil but Expensive Energy

Several gases contribute to the greenhouse effect, but carbon dioxide (CO_2) is the largest component, causing by most estimates about half of all global warming. "While this compound has always been a part of the earth's atmosphere, the burning of carbon-containing fuels has increased the atmospheric abundance of carbon dioxide by 25 percent over the past two centuries," according to Albritton.

Besides fossil fuel use, another, lesser factor contributing CO_2 to the atmosphere is the destruction of the world's forests. The photosynthetic process of trees consumes CO_2, but that process is lost as trees give way to civilization. Conversely, when trees decay or are burned, "fixed" carbon is released back into the atmosphere, creating CO_2.

Other major greenhouse gases include ozone, chlorofluorocarbons (CFCs), methane, and nitrogen oxides. Like CO_2, their introduction into the atmosphere is increasing steadily (see "Greenhouse Gases Other Than Carbon Dioxide" on page 25).

The vast majority of CO_2 is created by the burning of fossil fuels. There are two main options for reducing the CO_2 released during energy production: reducing demand by improving the efficiency of our energy use, or replacing "dirty" energy sources with safe alternative sources such as biomass, solar, hydro,

Greenhouse Gases Other Than Carbon Dioxide

Here's a rundown on the other greenhouse gases and their effect on global warming.

- Ozone forms a protective shield against harmful ultraviolet radiation when it is high in the stratosphere, but it is a dangerous toxic pollutant in the lower atmosphere and contributes to the blanket of warming gases covering the earth. Ozone is formed when nitrogen oxides in vehicle exhaust react with sunlight. Using more fuel-efficient vehicles, reducing driving, and switching to low-octane gas can reduce the creation of ozone.

- Chlorofluorocarbons (CFCs) and their chemical cousin halons are notorious primarily for their propensity to eat giant holes in the protective ozone layer. But they are also labeled as greenhouse gases; some types hold heat in the atmosphere 10,000 times more effectively than CO_2. They also stay in the atmosphere longer. The most common varieties, CFC-11 and CFC-12, have atmospheric life spans of 75 and 111 years, respectively.

The use of CFCs in manufacturing foam products such as Styrofoam containers and solvents accounts for 51 percent of CFC emissions, reports the Environmental Defense Fund (EDF) in its 1988 report *Protecting the Ozone Layer: What You Can Do*. EDF recommends using paper and plastic products instead of Styrofoam whenever possible, or foam products characterized by a bead texture (very little CFC is used for that kind of Styrofoam).

Another 45 percent of CFCs comes from car air conditioners and refrigeration systems. "Recovery and recycling of CFCs when products are discarded or serviced has the greatest potential for reducing CFC emissions," according to the EDF report. For example, it is standard practice when recharging car air conditioners to vent the old CFCs into the air. Recovery systems are available that clean and store the CFC for later use or to refill the air conditioner after it has been serviced, but appropriate disposal of CFCs is neither the law nor common work practice. The EDF recommends that policymakers institute "bounty systems" for recovery of CFCs from old cars and refrigeration systems, or deposit/refund systems in conjunction with recycling centers. In a recent major step, the major CFC-producing nations agreed to abide by the Montreal Protocol, an international agreement resulting from efforts of the United Nations Environment Programme. The protocol freezes the quantities of CFCs that signatory nations may produce and sets a schedule for eventual reductions, halving consumption within a decade. The U.S. regulations took effect on January 1, 1989.

Based on a new analysis issued after the Montreal agreement was reached, the U.S. Environmental Protection Agency (EPA) is expected to press for elimination of CFCs and halons by 1992 and a freeze on the use of methyl chloroform, an industrial solvent that also depletes the ozone layer. According to the EPA, the phase-down schedule for CFCs must be accelerated because the threat to the ozone layer is greater than previously projected.

- Methane is the "mystery" greenhouse gas because its sources are not well understood, nor is the extent to which it actually contributes to the problem. Scientists do know that methane is produced by agricultural activities, particularly the cultivation of rice, which emits methane as it grows, and the raising of cattle, which have a digestive process that creates methane. Methane is also produced in trash landfills as debris breaks down into its component chemicals. A handful of landfill operators recover methane to fuel electricity-generating plants. Environmentalists suggest reduction of waste and recycling as the best means of reducing the greenhouse effect of garbage.

- Nitrogen oxides are produced by burning of fossil fuels and the use of petroleum-based agriculture products that pollute both air and water. Jeremy Rifkin, president of the Foundation on Economic Trends, says that "Green Revolution" agriculture, which relies on artificial nitrogen fertilizers and pesticides made from petroleum products, must give way to sustainable agriculture based on organic farming.

From Public Citizen, *November/December, 1988.*

wind, or geothermal. Jeremy Rifkin, president of the Foundation on Economic Trends, says that to arrest the global warming trend, "we need to slash fossil fuel use by half by early in the next century."

With alternative fuels currently less available than cheap oil, energy efficiency stands out as an extremely important short-term goal. "We have to have [an alternative energy source] ready within a decade or two, but we have to get conservation in place first," says William Chandler, senior scientist at Pacific Northwest Labs, a subsidiary of the energy research and development firm, Batelle Memorial. Chandler sees efficiency as vital to development of renewable energy sources because an energy-efficient economy is much better suited to renewable sources. For example, solar collectors would have to be much larger to serve a building that uses its energy inefficiently than to serve an efficient building. "It is a sequencing problem. It is impossible to solve our problems with renewable sources without first implementing efficiency," he says.

Buildings use about one-third of the total energy consumed in industrial economies. Chandler recommends such simple steps as replacing large appliances with energy-efficient models and insulating the attic. A key to major increases in efficiency is "superinsulation," doubling the insulating capacity of a building and putting an airtight liner in the walls. Superinsulation adds 5 percent to the building costs but pays for itself in five years in saved energy and heating/cooling systems, according to the Worldwatch Institute, a nonprofit research organization. In a March 1988 report, *Building on Success: The Age of Energy Efficiency,* Worldwatch cites additional energy-efficiency strategies already being practiced to a limited extent: "smart buildings" that continuously monitor inside and outside temperatures to more efficiently heat or cool buildings; "thermal storage" technology, in which water cooled by night temperatures is used to air-condition buildings during the day; and integrated heating and cooling appliances. The report recommends a simple practice to relieve energy consumption by air conditioners: planting shade trees.

Buildings also lose heat and cool air through windows. "As much energy leaks through American windows every year as flows through the Alaskan pipeline," reports Worldwatch. Merely insulating windows during the winter can significantly lower the amount of energy used to heat a home or building. More advanced technologies may soon be available, including heat-reflective film that can be applied to window panes to double insulation capacity. Double-paned windows insulate like a Thermos bottle; their effectiveness may soon be improved by a transparent "aerogel," now under development, that can be injected between the panes. Different combinations of these techniques may soon yield windows that insulate as well as walls.

Another area of great potential efficiency is lighting. Fluorescent bulbs that can replace the incandescent ones currently used in most household lamps are starting to appear on the market. In a 1988 report, *Energy Efficiency: A New Agenda,* the American Council for an Energy Efficient Economy (ACEEE), says that one 18-watt fluorescent bulb provides the light of a 75-watt incandescent, lasts ten times as long, and over its useful life will keep 130 kilograms of carbon out of the air when compared to the less efficient conventional bulbs. Fluorescent bulbs are more expensive, but the savings in energy and longevity of the bulb easily offset the purchase price. Unfortunately, highly efficient bulbs are difficult to find since most stores won't carry them because of the price, reports Chandler. Consumer pressure on hardware and building supply stores will be necessary to change this situation.

Utilities are well positioned to educate and encourage their customers to be more energy-efficient. One way that some states have found to make utilities serve that function is by imposing least-cost planning policies. Under least-cost planning, utilities that want to expand their capacity must, whenever economical, invest in efficiency improvements instead of new power sources. For example, utilities can offer consumers energy audits and incentives for them to invest in efficient heaters and appliances. These types of measures can save as much energy across the customer base as would be generated by new power plant capacity. In 1987, statistics compiled by the Energy Conservation Coalition, 23 state regulatory commissions have implemented or are considering least-cost planning.

Another approach to promoting efficiency links plans for expanded capacity to environmental effects. Joseph Goffman, senior attorney for the Environmental Defense Fund (EDF), says one idea that federal policymakers should consider is to require utilities that are planning to expand capacity to reduce their total CO_2 emissions by the amount that would be produced by adding any new capacity. This approach would not ultimately reduce CO_2 production, but it would also slow any increase and would force utilities to invest in efficiency technology. "The most important part of this," says Goffman, "is that we would be building experience with efficiency techniques."

Energy analysts like Chandler say that efficiency measures such as those described above offer the most potential for CO_2 reduction. Efficiency is a proven commodity, and it is available now. It is also a strategy with which the U.S. and other industrial countries have experienced great success. According to the ACEEE, the U.S. economy has grown 40 percent since 1973, when the oil crisis triggered a national effort to conserve, but energy use remained steady through 1987. In addition, the environmental benefits of energy efficiency are underscored by the economic benefits. In its 1988 report, the ACEEE notes that if the U.S. made a $300 million to $500 million investment in energy efficiency, some $1.3 trillion to $2.2 trillion spent on oil imports (in 1987 dollars) could be saved over the next 20 years.

Cutting Down on Car Carbon

Energy efficiency is also the key to reducing the significant contribution that motor vehicle emissions make to global warming. The Natural Resources Defense Council reports that if the U.S. auto fleet achieved an average of 45 miles per gallon by the year 2005, U.S. CO_2 emissions would be reduced by about 15 percent. Some analysts think it is possible to achieve overall fleet efficiency of 60 miles per gallon with current technology.

Reinstituting the 55-mile-per-hour speed limit is another way to reduce combustion and the polluting by-products, according to Gordon J. MacDonald, vice president and chief scientist of the MITRE Corporation, in testimony at a January 1987 congressional hearing on the greenhouse effect. "Enforcement of speed limits is probably the most cost-effective means of encouraging conservation in the transportation sector," he says.

In the long term, better urban planning could incorporate mass transit and also make commercial buildings more conveniently located, to reduce the overall amount of driving Americans do. While many cities and suburbs are not amenable to drastic reordering, future urban planning decisions must make it a goal to reduce America's reliance on the automobile.

"Both a carrot and a stick are required to move societies toward more energy-wise transport systems," says the Worldwatch Institute in its *Building on Success* report. "A first priority might be to eliminate some of the hidden subsidies to automobile travel. Drivers almost never pay the full cost of their cars, including road building and maintenance, traffic congestion, noise, air pollution, and danger to pedestrians. In many cities, traffic congestion makes traveling hectic and time consuming, exacting millions of dollars in uncounted economic losses."

Efficiency Alone Can't Save Us

Though energy efficiency has vast potential to reduce current CO_2 emissions, EDF's Goffman notes, "There is only so much slowing of CO_2 buildup that efficiency will get you. Nonfossil fuels are eventually needed." The only way to be able to make the transition from a fossil-fuel-dependent economy, he ex-

---

Earth Care Action

Ten Steps to Turn Off the Heat

By Curtis A. Moore

Used in combination, the following steps could take the world from the abyss of environmental despair to a realistic Promised Land, with virtually zero pollution. And it could do so by 2050, or even earlier, using existing technologies and actions—if only we have the will.

Step One: Switch to natural gas where possible. It cuts all three of the critical air pollutants, halving carbon dioxide, the main greenhouse gas. When natural gas is burned in one of the new super turbines already available, air pollution is reduced still further.

Step Two: Burn coal more efficiently. Switching to a new technology such as pressurized fluidized-bed combustion cuts sulfur pollution by 94 percent and oxides of nitrogen by 30 percent over conventional coal-burning power-plant operations.

Step Three: Cleanse the stack gases. Pollution controls added on to existing coal-fired power plants cut sulfur dioxides and oxides of nitrogen by up to another 90 percent. One technology has already been installed on 400 power plants abroad.

Step Four: Don't waste the heat. Instead of burning fuel in one place to generate electricity and another to manufacture products, combine them. Cogeneration boosts total efficiency to 85 or 90 percent, dropping air pollution by another 40 to 50 percent.

Step Five: Build better cars. In the United States alone, a five-mile-per-gallon improvement in auto mileage would cut carbon dioxide pollution nearly 200 billion pounds a year. Fuel-efficient cars would cut automotive carbon dioxide by up to 70 percent.

Step Six: Phase out freons. Chlorofluorocarbons (CFCs) could be virtually eliminated in five years if governments mandated that existing alternatives be used. An additional ban on CFC spray-can use would cut this gas by 25 percent.

Step Seven: Make homes more efficient. Replacing the current generation of energy-hungry water heaters, air conditioners, furnaces, and light bulbs would cut household energy consumption—and air pollution from homes by 50 to 66 percent.

Step Eight: Halt tropical deforestation. New schemes that allow multiple harvesting, sustainable development, and replanting of certain species—instead of burning virgin jungle—can cut air pollution from a given acre by 90 percent.

Step Nine: Curtail industry energy consumption. Industries in the U.S. and Canada gobble twice as much energy per unit of production as their Japanese counterparts. Put them on a par and industrial air pollution is cut by 50 percent in North America.

Step Ten: Use hydrogen fuel. Over the long term, development of radically different technologies, including those that use hydrogen, is the answer, since interim phase-out of dirty cars and power plants may be offset by population and economic growth.

From International Wildlife, *May/June 1989. Reprinted by permission.*

plains, is to invest now in the development of renewable energy sources that are cost-effective as well as clean.

Small hydroelectric plants, wind turbines, biomass, geothermal plants, photovoltaic cells, and active and passive solar heating systems are categories of renewable energy technologies that are available commercially and that already provide 8.6 percent of the U.S. energy supply, according to *Turning Down the Heat: Solutions to Global Warming*, a fall 1988 report by the Public Citizen Critical Mass Energy Project and the Safe Energy Communication Council. The report says that a "cost-effective hybrid" of energy-efficiency improvements, renewable energy resources, and increased use of natural gas can cut carbon dioxide emissions produced in the industrial, residential/commercial, and utility sectors by 68 percent by the year 2000.

Other energy options such as cogeneration and combined-cycle natural gas turbines currently produce electricity more cheaply while producing far less carbon dioxide than do coal-fired plants, according to the report. These energy sources can provide a "transition bridge" for electrical generation during the remainder of this century while producing less than half the volume of greenhouse gases caused by burning coal and oil. Critical Mass and other safe energy advocates hold the position that safety factors, economic considerations, and others preclude nuclear power from being a reasonable "safe energy" alternative to fossil fuels.

Deforestation:
A Double-Edged Disaster

The destruction of the world's forests to make way for humans and agriculture is a devastating contributor to the increase of carbon in the earth's atmosphere. Growing forests have a net effect of reducing CO_2 because they use up more of the chemical for growth than they release through decay in the forest. Mature forests are the storage closets for immense amounts of carbon, which is released into the atmosphere when they decay or are burned.

John Robbins, author of the 1987 book *Diet for a New America*, which advocates vegetarianism on environmental and economic as well as moral grounds, maintains that the voracious demand for cheap beef in the U.S. is a primary force behind deforestation in the Western Hemisphere. Cattle production is subsidized by the U.S. government in a variety of ways, including cheap leases on government land. Even cheaper economics for raising cattle in Central and South America means that 10 percent of beef consumed in the U.S. is now imported. Cattle ranching has become a major source of income in the Americas; with land concentrated in the hands of a few, ranchers and farmers are expanding into the jungle. According to an August 1988 *New York Times* article, scientists suspect that the immense pace at which Brazilian forests are being slashed and burned may account for at least one-tenth of the global man-made output of carbon dioxide. That estimate doesn't factor in the effect of the forest destruction taking place elsewhere in the world.

A necessary step the United States must take is to institute national policies that track overall forest acreage, maximize the foresting of federal lands, provide for the export of least-cost energy and energy-efficiency technologies to Third World countries to lessen their dependence on wood burning for energy, and create financial incentives for countries to preserve their forests. Pressure must also be brought on international lending institutions to ensure that the development projects they underwrite are ecologically sound and do not unnecessarily destroy woodlands. One method being tried by various environmental organizations to discourage forest devastation in other countries is the debt-swap, where the quid pro quo for debt relief for developing countries is establishment of parks in the rainforests.

Direction from the Top

Despite the new urgency that the global warming threat places on energy efficiency, the U.S. is backsliding. After remaining steady for many years, U.S. energy use increased at a 7 percent annual rate during the first half of 1988, according to Chandler at Pacific Northwest Labs.

The strongest determinant of energy use is price, and high energy consumption has been driven by the collapse in oil prices. Oil prices fell by 75 percent between 1981 and early 1988, according to the Department of Energy. "With oil prices so low," says Goffman, "we need policies that will help create demand for efficiency."

The Reagan administration has demonstrated antagonism toward conservation and clean energy issues, however. For example, the U.S. embarked on a path toward fuel efficiency in 1975 with the Energy Policy and Conservation Act, which required automakers to achieve a fleet average of 27.5 miles per gallon by 1985. These standards have been relaxed for cars of the 1986 model year and beyond, rather than strengthened, by the Reagan administration. In addition, research and development into energy-efficiency technology has been cut by about 55 percent during the Reagan years, according to ACEEE, and renewable R&D has been slashed by closer to 80 percent.

The United States must take the lead in actions to curb global warming not only because this country has the greatest resources to do so, but because it creates an inordinate share of the problem. The World Resources Institute calculates that 47 percent of worldwide CO_2 emissions come from the U.S. and the Soviet Union, while all of the developing nations together contribute only 20 percent. According to Michael Totten, legislative assistant to Rep. Claudine Schneider (R-R.I.), "People born in the U.S. between now and the year 2000 will give off more carbon from burning fossil fuels than everyone born [in the same time period] in all of Latin America and Africa."

Rep. Schneider is one of several federal lawmakers who are introducing legislation to address these issues. In October, she introduced "The Global Warming Prevention Act," a bill that attempts to define the path to a transition to renewable energy sources in 10 to 30 years. In addition, it addresses forest and agriculture policies, housing policy, development assistance, waste production, and population growth in an effort to reduce the potential of each of these activities to add to global warming.

Sen. Timothy Wirth (D-Colo.) has introduced legislation that is similar but also includes $500 million for development of "inherently safe and cheaper" nuclear power. *Public Citizen* has criticized that aspect of Wirth's bill as being "unnecessary" to solve the problem. Sen. Robert Stafford (R-Vt.) also has introduced legislation that establishes targets and timetables for reducing greenhouse gases but has fewer specific policy proposals.

The attractive irony to the global warming crisis is that the measures necessary to stop the trend provide other benefits as well: reduced land and air pollution, savings in energy costs, and preservation of forests and tropics. "I am mildly optimistic," says Rifkin, "if we can take an ecumenical approach that crosses constituencies and political boundaries. It is a matter of rolling up our sleeves."

Earth Care Action

Solutions for an Overheating World

Slowing global warming will require action on many fronts. This article highlighting key strategies was prepared by *Environmental Action* and Environmental Action Foundation staff working on these issues.

The Most Promising: Energy Efficiency

By Nicholas A. Fedoruk

Energy efficiency offers the most immediate and the most substantial gains in our drive to cut greenhouse gases. Almost any action that makes energy use more efficient will mean less carbon-based fuels are burned, whether those efficiency measures save electricity, oil, or other fossil fuels, whether they are implemented in buildings, power plants, homes, or cars.

In the past 15 years, U.S. efficiency investments have cut energy consumption by one-third and carbon emissions by 40 percent. Public and private analyses demonstrate that the United States could double the resulting savings of $160 billion per year while maintaining vigorous economic growth. By comparison, investment in natural gas (which emits less CO_2 than coal or oil), fossil fuel, and nuclear technologies could not have a comparable impact before 2000 and would cost two to five times as much.

Below, two priority areas are described; other important efforts include programs that improve mass transit and encourage its use: more energy-efficient housing for all incomes, expanded appliance efficiency, and more efficient manufacturing processes.

Less gas per car. Automobiles and light trucks account for 20 percent of all U.S. carbon dioxide emissions, the major greenhouse gas.

A car that gets only 18 miles per gallon—the current average for cars driving U.S. roads—will produce more than 57 tons of CO_2 during its lifetime. A car that gets 26.5 miles per gallon (the standard set for automakers' new 1989 fleet) will emit 20 tons less.

Even more carbon emissions can be kept out of our saturated atmosphere if Congress sets firm fuel-efficiency standards. (Doing so will also reduce consumption of the world's dwindling oil supply and cut car-owners' operating costs.)

New vehicles should be required to get 45 miles per gallon by 2000 (lifetime emissions of 26 tons of CO_2) and 60 miles per gallon by 2010 (lifetime emissions of 17 tons). These standards are technologically feasible today. The Geo, now appearing in dealer showrooms, gets 58 miles per gallon highway, 53 city. And prototype cars are ready for production that achieve 80 miles per gallon.

Electricity with efficiency. Three-quarters of U.S. electricity is generated by burning those global warming agents—fossil fuels. The potential for reducing that combustion through energy conservation is tremendous. The utility industry's main research body, Electric Power Research Institute, has estimated that just applying current technology could reduce peak demand 25 percent by the year 2000.

Duke Power in North Carolina and other utilities are introducing "appliance control" and industrial "load-shedding" practices. Customers are encouraged to buy energy-efficient appliances, weatherize their homes, and replace inefficient lighting, heating, cooling, and hot water systems. Commercial customers are offered incentives to install high-efficiency lighting equipment. Load-shedding is a combination of shifting the time of day that electricity is consumed and cutting off equipment during periods of high demand. The more advanced utility efforts offer actual financial incentives (rebates, loans, and grants) for customers large and small to participate.

Duke Power projects that by the year 2000, such efficiency measures can reduce peak load demand by 6.900 megawatts, equal to six or seven large power plants.

Reducing Air Pollution Brings a "Two-Fer"

By Daniel Becker

Many sources of air pollution pose threats not just to public health, forests, lakes, and streams. They also add to the thermal blanket of greenhouse warming. Congress has been grappling (so far unsuccessfully) with versions of the Clean Air Act that would control each of these. Passage would yield a "two-fer," two benefits for one cleanup.

- Ozone smog is a greenhouse gas and it makes breathing the air of cities across the country an unhealthy proposition. Ozone smog forms when hydrocarbons and NOx (short for nitrogen oxides, which include NO and NO_2) mix in strong sunlight. These ozone precursors come from automotive, industrial, and power-plant emissions.

- Electric power plants are major sources of acid-rain-causing sulfur dioxide and NOx. They also pump out billions of tons of CO_2, the most plentiful greenhouse gas, as well as substantial nitrous oxide.

- Chlorofluorocarbons (CFCs) are destroying the earth's stratospheric ozone shield and are also among the most potent greenhouse gases.

To address acid rain, utilities, vehicles, and industrial sources must reduce emissions of nitrogen oxides by 4 million tons a year and sulfur dioxide by 12 million tons a year. To control ozone smog, cities need a strong federal program and tighter auto emissions standards. To control toxic air pollutants, we must require the Environmental Protection Agency to run an effective program. And Congress should enact a complete ban of CFCs.

Yet Another Reason to Recycle

By Jeanne Wirka

During the 1970s, recycling picked up momentum as a resource and energy conservation measure to combat the energy crisis. During the 1980s, it was embraced as a solution to the nation's landfill crisis. Now, as we approach the 1990s, recycling can be seen as part of a strategy to reduce global warming.

In manufacturing, energy is saved whenever secondary (recycled) materials replace virgin materials. Less energy then goes to extracting raw materials, processing them, and transporting them. How much energy can be saved?

An Environmental Protection Agency study predicts that in the year 2000, the United States will landfill or incinerate 11.4 million tons of newsprint, 16.2 million tons of corrugated cardboard, 10.8 million tons of glass packing, 8.2 million tons of plastic packaging, and 1.5 million tons of aluminum packaging.

If these 48.1 million tons were recycled instead of tossed out, the nation would save the energy equivalent of 10.1 billion gallons of gasoline, according to a recent study by the National Appropriate Technology Assistance Service (NATAS). Decreased energy use will reduce greenhouse gas emissions. The NATAS study reports that by using recycled instead of virgin materials, manufacturing processes can cut general air pollution by up to 22 percent for glass, 73 percent for paper, and 95 percent for aluminum.

Additionally, recycling can help avoid a new global warming threat: Garbage-burning incinerators.

Incinerators emit CO_2, just as any other combustion process does. In fact, current incinerator technology aims to maximize CO_2 emissions, because this means a "more efficient" burn with less carbon monoxide (CO)—which is, unlike CO_2, a regulated air pollutant. Incinerators also spew out a whole range of other pollutants, some of which (such as nitrogen oxides) may contribute to the formation of greenhouse gases. A sound recycling program can either eliminate a city's need for an incinerator or allow a contraction in its planned size.

Trees: The CO_2 Storers

By Drusilla Schmidt-Perkins

Trees can help offset—or worsen—global warming. As a tree grows, it absorbs CO_2 at a rate that varies with species and age. Faster-growing trees consume more, but longer-lived species "fix" the gas for longer periods. The American Forestry Association uses the benchmark average of 13 pounds of CO_2 per year per tree.

When trees are cut down, the carbon they've absorbed is released—at a slow pace if the tree decays naturally and instantly if the tree burns.

Deforestation contributes to global warming in two ways. Burning of clear-cut trees releases CO_2, and those trees are no longer available to absorb CO_2. Tropical forests in the early 1980s were being cleared at the rate of about 28 million acres a year. Halting Third World loss of rainforests can help slow global

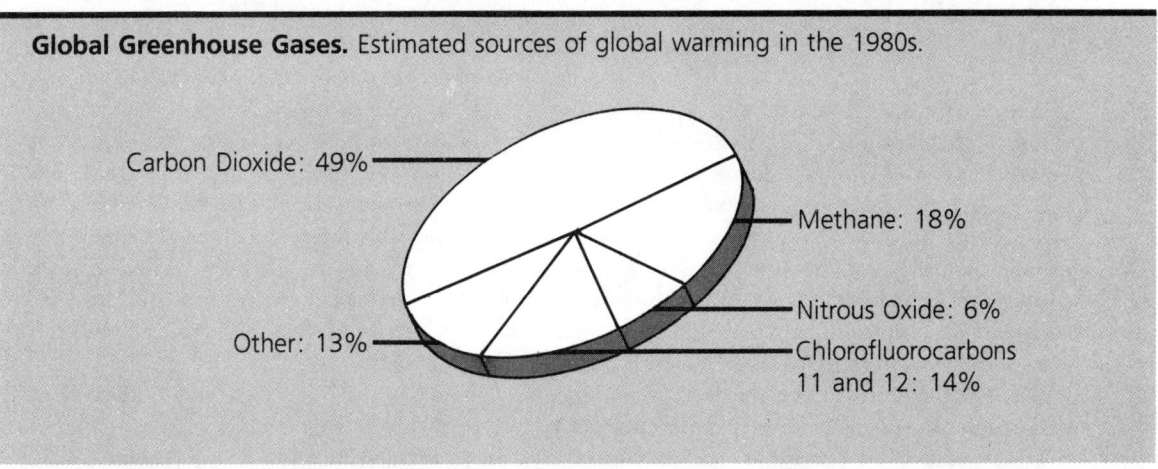

Global Greenhouse Gases. Estimated sources of global warming in the 1980s.

- Carbon Dioxide: 49%
- Methane: 18%
- Nitrous Oxide: 6%
- Chlorofluorocarbons 11 and 12: 14%
- Other: 13%

SOURCE: Based on an article in *Journal of Geophysical Research*, by James Hansen, August 20, 1988, as charted by Renew America.

warming. So will planting trees anywhere on the earth.

Besides directly removing CO_2 from the atmosphere, trees can help conserve energy. Three well-placed trees around a home can reduce air-conditioning needs by 10 to 50 percent; this in turn means the local utility can burn less fossil fuel.

The same holds true for cities. Cities are often 10 degrees warmer than suburbs, partially due to the "heat island effect" caused by urban agglomerations of concrete, steel, and asphalt. Because trees can minimize this phenomenon, utilities are beginning to fund city tree programs under energy-efficiency programs.

Planting trees won't "solve" global warming, however. To offset current levels of global CO_2 output, some three billion acres of trees would need to be planted worldwide for each year of emissions—an area more than one-and-a-half times the continental United States.

The Global Warming Agenda for Mr. Bush

The President should announce that minimizing global warming will be a top priority of his domestic and foreign policy. He should issue an executive order establishing goals and defining the responsibilities of all relevant agencies. He should work with the Congress to develop and enact appropriate legislation.

The President should direct the Secretary of State to make bilateral approaches to key nations, including the Soviet Union and major developing countries like China and India, and to work with other nations to develop a global treaty requiring that CO_2 emissions be reduced through increases in energy efficiency and greater reliance on renewable energy sources.

The President should propose that other nations join us in a major program to halt tropical deforestation and to plant trees on a massive scale.

The President should instruct the Environmental Protection Agency (EPA) or take all actions necessary to phase out U.S. use of chlorofluorocarbons in five to seven years. The President should direct the State Department and EPA to work with the United Nations Environment Programme and other nations to strengthen the international ozone depletion protocol signed in Montreal in September 1987.

What Congress Can Do

By Nicholas Fedoruk and Daniel Becker

Three broad-ranging bills to address the greenhouse threat were introduced in the last session of Congress, setting the stage for action in the new term. Each is likely to reenter the fray in 1989, although in somewhat different form. Other legislators will draft narrower bills that compensate for their more circumscribed attack on the problem with what many observers cite as an improved chance of passage.

One of the comprehensive bills sets an ambitious target for cutting back greenhouse gases. Introduced in July by Sen. Robert Stafford (R-Vt.) and Sen. Max Baucus (D-Mont.), the bill sets the goal of a 50 percent phase-out of CO_2 emissions by 2000, bans CFCs as of 1999, and calls for cutbacks in nitrogen oxide and hydrocarbon emissions. Reduction requirements are set for electric utilities, vehicles, industries, and residences, although no penalties are spelled out.

The other two bills (Schneider and Wirth-Johnston) are more prescriptive. Both concentrate on mandating and funding specific energy programs that can reduce greenhouse gases.

Both include a host of other programs to encourage reforestation, ecologically sound development assistance, global population planning, and international cooperation on global warming. The two bills—one proposed in July by Sen. Tim Wirth (D-Colo.) and Sen. J. Bennett Johnston (D-La.) and the other in October by Rep. Claudine Schneider (R-R.I.)—establish basically identical goals of a 20 percent reduction in CO_2 emissions by the turn of the century. Among the energy programs:

- Each mandates writing of a "least-cost" national energy plan to guide federal and private energy use. The Schneider bill adds $500,000 for public participation, and provisions that promote least-cost planning by electric utilities and foreign aid recipients.

- Each funds about $800 million worth of federal research into energy efficiency over several years and about $500 million worth of research and development in renewable energy.

- Both bills call for roughly 60 percent higher mile-per-gallon ratings by the year 2000 for cars and light trucks. The Schneider bill adds an increase in the gas guzzler tax, rebates for buying highly efficient vehicles, and requirements on how fuel-efficient federal fleets of vehicles must be.

In most sections, the Schneider bill more aggressively assists energy efficiency and renewable energy. But another aspect definitively sets the bills apart.

The Wirth-Johnston bill gives a boost to the moribund nuclear industry by authorizing $500 million for research into a new generation of "passively safe and cost-effective nuclear power plants." Most of the money would go to construct a demonstration reactor. Environmental Action and other groups have announced that they will oppose the bill as long the nuclear power provision remains.

What You Can Do

Think of our environment as the proverbial camel with a straining back and our daily lives as filled with the unseen addition of slender straws to the beast's ever-growing burden. Obviously, our individual contribution to the load is small. But changing our behavior can remove a few straws—and maybe set an example for others. Cumulatively, changing human behavior is the only way to save our poor camel's spine.

Make energy-efficient choices whenever possible. Buy energy-efficient appliances. Bicycle or use public transportation. Drive a car that gets good gas mileage—and drive only when necessary.

Keep your furnace in good repair. Some 18 percent of U.S. CO_2 emissions comes from home systems that burn fossil fuels for cooking, heating, hot water, etc. (rather than using electricity). Because they tend to put out more CO_2 than gas-fired furnaces, an oil-fired system must be thoroughly cleaned and tuned yearly. This should include vacuuming the heat-exchanging surface, replacing the burner nozzle (which should be as small as possible), and a thorough efficiency test—CO_2 percentage should be 11).

Recycle. Keep valuable natural resources from being burned or landfilled by recycling. Products made from recycled materials require less total energy in their manufacture.

Choose ecological products. Avoid non-recyclable packaging and plastic foams made with CFCs.

Plant trees. Each will take some CO_2 out of our overheating atmosphere.

Become a greenhouse activist. Join environmental organizations to further the group's activism. Use your power as a citizen by writing to Congress, the President, your state, and your utility.

What States Can Do

By Sheila Machado and Rick Piltz

States can play an important role in reducing the rate of global warming by providing models of effective programs to combat the greenhouse effect. This doesn't negate, however, the need for a strong federal presence to ensure that all states implement effective policies.

• At least ten states have substantial statutory authority and policies requiring that utilities use least-cost planning and investment practices.

• Fifteen states offer alternative energy tax credits.

• Sixteen states fund ride-sharing programs. Nine states have maintained the 55-mile-per-hour speed limit. Nine states have high-occupancy-vehicle lanes.

• Four states—Maine, Vermont, Massachusetts, and Rhode Island—ban or restrict Styrofoam products, some of which contain CFCs. States could take the lead in using regulations or economic incentives to promote CFC recovery and recycling, especially from refrigerators and auto air conditioners.

• To combat ozone smog, 4 states have statewide inspection and maintenance programs for motor vehicle emission systems, and 29 states have city- or county-wide programs.

Between Nations

By Rose Marie L. Audette

Several diplomatic agreements tackling the greenhouse phenomenon or aspects of it are in the works; others have been concluded.

An Energy-Hungry Nation. Americans consume considerably more than their share of the world's energy.

World Population — United States: 5%

World Energy Consumption — United States: 29%

SOURCE: Environmental Protection Agency

Stronger and more far-reaching international action is essential, however, to improve the world's prospects for avoiding climate catastrophe.

Global warming treaty. The United Nations Environment Programme has established a special Task Force on Climate Change and targeted 1995 for completion of an agreement on global warming measures.

Helping to forge the way for such a treaty is the "Toronto Conference," which took place in June 1988. The gathering of prominent government, corporate, and environmental leaders from 48 countries, although without official status, recommended a 20 percent reduction in worldwide use of fossil fuels by 2005 and an eventual reduction by 50 percent. In addition to enhanced energy efficiency, participants called for a switch to lower CO_2-emitting fuels, emphasis on renewable energy, and "revisiting the nuclear power option." The conference also urged creation of a "World Atmosphere Fund," to be financed by a tax on fossil fuel consumption in industrialized nations.

CFC treaty. Under the Montreal Protocol, signatory nations have agreed to reduce production of ozone-destroying and globe-warming CFCs by 50 percent by 1999. A complete ban on CFCs is the next step.

NOx and SO_2 agreements. Sources of these acid rain precursors emit other greenhouse gases, and NOx helps form ground-level ozone, also a greenhouse gas. The "30 Percent Club" has agreed to cut SO_2 emissions by 30 percent; the U.S. has refused to join its 21 member nations. In late October, 12 Western European nations agreed to cut NOx emissions by 30 percent over ten years. The next day, a larger group that includes the United States signed the Nitrogen-Oxides Protocol, a much weaker agreement that freezes NOx emissions at the 1987 level, beginning in the mid-1990s.

Who pays? In October, the "Global Greenhouse Network" was founded during a Washington conference of activists (including environmentalists, church leaders, Third World representatives, and elected officials) from 35 nations.

The informal coalition expressed concern "that the burden of change remain primarily with the Northern industrialized nations that are chiefly responsible for the emissions of greenhouse gases and not be shifted to the poor of the Third World."

The network also advocates retiring Third World debt as an incentive to conserve tropical forests and a shift from chemical pesticides and fertilizers to "organic-based ecologically sound agriculture."

Reprinted from Environmental Action *magazine, January/February 1989. Reprinted by permission.*

Reactors Redux

By Michael Philips

It's an exemplary promo plan. The music ebbs and flows in the background; the narrator's voice massages his unseen audience. The topic is power without fear. General Atomics, we are told, has a new process for generating electricity, a process that will satisfy the public, the utilities, and the investment community. The product—the high-temperature, gas-cooled nuclear reactor—may sound technical and recondite, but it lays claim to some comforting and simple adjectives: safe, reliable, and cost-effective.

1953? No, 1989. The General Atomics nuclear reactor is just one of several so-called advanced models emerging in the energy spotlight as concern over global warming increases. To more and more policymakers, the newly designed reactors are feasible alternatives to coal, oil, and natural gas power plants, all of which emit carbon dioxide (CO_2), one of the primary culprits behind the greenhouse effect.

Somewhat fewer people see the reactors as preferable to the large, expensive, and cumbersome water-cooled models that, despite their faults, have so far been the United States' reactors of choice. With sufficient federal support, the gas-cooled version could, its sponsors speculate, be in the vanguard of a second generation of nuclear reactors.

Even some members of Congress who have historically made themselves at home in the environmentalist camp are tempted by the new reactors, saying that in light of the greenhouse effect, it's time to reevaluate nuclear power.

Any such appraisal must begin with an understanding of the contribution nuclear reactors could reasonably make toward reducing greenhouse gas emissions. Obviously, to the extent that a nuclear plant is built instead of, or as a replacement for, an existing fossil-fuel plant. CO_2 emissions will be reduced. If it were possible to replace all U.S. fossil-fuel-burning plants with nuclear facilities, total CO_2 emissions would be reduced by 28 percent nationwide. (This estimate does not reflect the undetermined but possibly significant amount of fossil fuels used—and CO_2 emitted—during the construction and operation of a nuclear plant.)

But CO_2 is not the only malefactor. Taking into account the methane, chlorofluorocarbons, and other trace gases that are also warming the planet, nuclear power could offset only 14 percent of all U.S. greenhouse gas emissions.

Nuclear power's contribution dwindles as the territory expands. Replacing all U.S. fossil-fuel plants with nuclear reactors would reduce global greenhouse gas emissions by about 4 percent. And as the use of fossil fuels grows in the Third World, the figure drops further.

Assuming a full-scale nuclear mobilization were nonetheless desirable, and assuming no inflation in the 1987 average cost of constructing a nuclear reactor in the United States, replacing the nation's fossil-fuel-generated electricity with nuclear-generated electricity would cost $1.2 trillion. And if nuclear advocates are not yet deterred, there's one more consideration: According to analyst Bill Keepin of the Rocky Mountain Institute, to replace coal-burning power plants, nuclear capacity would have to increase "at the staggering rate of one large nuclear plant every 1.6 days for the next 38 years" in a high-growth scenario and every 2.4 days in a medium-growth scenario. Utility analyst Charles Komanoff adds that nuclear reactors would have to come on-line worldwide at the rate of eight per week to displace just half the CO_2 emissions of fossil-fuel plants over the next 35 years.

Formidable as these numbers are, they do not daunt the nuclear industry. In press releases and congressional testimony, industry representatives repeatedly tout nuclear as CO_2-free. And when critics pull out the numbers showing nuclear's limited potential, industry spokespersons hold up the red flag of global warming.

Indeed, one of the major pieces of federal legislation addressing the greenhouse effect, introduced by Sen. Timothy Wirth (D-Colo.), proposes, among other things, spending $500 million to demonstrate the feasibility of advanced reactors. Wirth, considered one of the more environmentally minded members of Congress, has said that Americans must get over their "nuclear measles" and that "environmentalists will come around. They can't help but come around."

Perhaps more to the point is whether, given its technical, financial, and regulatory difficulties, nuclear power itself can come around. No new reactors have been ordered in the United States since 1978, and all those ordered between 1974 and 1978 have been canceled. The problems at the 111 licensed civilian reactors make it unlikely that any new ones will be ordered soon. Many facilities are aging prematurely, their reactor vessels becoming "embrittled"—that is, weakened from continuous exposure to radiation. Despite modifications, the eight reactors designed by Babcock & Wilcox (manufacturers of the reactors at Three Mile Island) are still trouble-prone, while the containment systems for 24 operating General Electric reactors are, according to a study sponsored by the Nuclear Regulatory Commission (NRC), likely to fail in the event of a major accident.

After more than three decades of commercial operation, nuclear power has yet to prove itself safe. During the last decade, U.S. nuclear utilities have reported nearly 30,000 mishaps at their plants, including the partial meltdown in 1979 at Three Mile Island and several close calls since then.

The NRC, the agency charged with oversight of the nation's nuclear power plants, inspires little confidence. Congressional hearings have focused on the NRC's "coziness" with the industry it regulates: Rule changes often benefit the industry; safety procedures are haphazardly enforced; and in general, the agency is unwilling to require safety improvements that might be expensive for a utility to implement. The NRC is, in short, reluctant to take any action that might imply that all is not quiet on the nuclear front.

The present aside, no one yet knows what to do with a reactor once it has reached the end of its useful life. Nor has a politically and technologically trustworthy decision been made regarding nuclear waste. Last year [1988] Congress decided to send the nation's high-level nuclear waste to an underground repository to be built below Yucca Mountain, Nevada, but controversy over that choice has delayed exploratory work, and few observers expect Yucca Mountain to open its doors on schedule in 2003.

New Designs, New Promises

Advocates of advanced reactors believe that the new design of their machines will diminish concerns about nuclear safety and, by extension, about regulation. Senator Wirth's interest in nuclear power as part of any greenhouse solution is predicated on the belief that an "inherently safe" reactor can in fact be built.

Suddenly the term "inherently safe" is everywhere. Sometimes the somewhat less colorful term "passively safe" is used; one advanced-reactor design is even named PIUS—Passive Inherent Ultimate Safe.

For those who believe nuclear's woes are a result of bad P.R. and an irrationally fearful public, the new and improved lingo may hold a certain attraction. But Robert Pollard, a former safety engineer with the NRC who now works with the Union of Concerned Scientists, points out that fissioning atoms is, if anything, inherently dangerous. You can compensate for that and perhaps make it acceptable to the public, he says, but you can't make it inherently safe.

Many in the nuclear industry itself are also uncomfortable with the terminology. "The connotation of the term is misleading," says Carl Goldstein of the U.S. Committee on Energy Awareness, a nuclear-industry trade group. "We absolutely cannot tell the public that something is inherently safe." Goldstein says a significant number of his organization's members feel likewise.

Semantics aside, how advanced are the advanced reactors? While there have been the sorts of improvements in circuitry, piping, and so forth that one would expect in any technology, the basic advance is in the method of cooling the reactor core. All nuclear reactors must have a system for cooling the core in case of an accident, to avoid melting the reactor fuel and releasing radioactivity into the environment. The new designs all move away from the conventional "active" pumping systems laden with moving parts to "passive" systems that rely to a greater extent on natural forces such as gravity.

The advanced reactors sound impressive. Not only are they supposed to be safer, they'll be smaller and possibly less expensive than conventional nuclear plants. There's just one problem: Most exist only on paper. Karl Stahlkopf, who directs advanced-reactor research at the Electric Power Research Institute in Palo Alto, California, notes that while portions of the designs do exist at some plants in the United States and overseas, "We're looking well into the next decade before any of these are certified" by the NRC.

General Atomics disagrees. The company believes that the basic concept of the high-temperature, gas-cooled reactor (HTGR) has already been proved and that only the specifics need to be demonstrated. The company also says that while it is confident of the HTGR's cost-effectiveness, it does need several hundred million dollars from the federal government to prove this fact. William Moomaw, director of the World Resources Institute's climate and energy program, questions the wisdom of such a subsidy: "With an industry as mature as the nuclear industry, shouldn't it be paying for this itself?" Pollard of the Union of Concerned Scientists concurs: "If you can't find private investors, that should tell you something."

Ultimately, the safety assurances of General Atomics and others melt away when the companies are pressed on the issue of accident liability. The Price-Anderson Act, renewed by Congress last year, protects the nuclear utilities—and by extension the reactor manufacturers—by placing a dollar limit on the liability they would face in the event of a nuclear accident. Is the industry willing to forgo that coverage and accept financial liability for an accident at one of their "inherently safe" reactors? "No," says Tom Johnston, executive consultant to General Atomics. But if the reactors will be so safe, then why not? "People in this business," Johnston explains, "are very, very conservative."

A Far Better Solution

Despite attempts by the nuclear industry to characterize environmentalists as being "split" on the question of a nuclear revival, the environmental community generally agrees that nuclear power need not play a role in any comprehensive response to global warming.

The best CO_2-free "source" is energy efficiency. Analyses conducted at Princeton University and the Lawrence Berkeley Laboratory show that developed countries could cut per capita energy demand by as much as 50 percent over the next 20 to 30 years and still maintain current rates of industrial expansion and economic growth.

Although improvements in energy efficiency will require capital investment, Rocky Mountain Institute's Keepin estimates that each dollar put into efficiency displaces seven times more CO_2 than a dollar spent on nuclear power.

The United States in particular has a long way to go to improve its energy efficiency. Last summer, the World Resources Institute's Moomaw testified before the Senate that "Japan produces a dollar of GNP, or the equivalent in yen, while using about half of the energy" that Americans use.

After efficiency, renewable sources of energy such as solar, wind, and geothermal power are the next-best energy investments. According to the Department of Energy, renewable resources constitute one of the nation's largest energy reserves—more than five times the energy potential embedded in the country's coal reserves—and could meet up to 80 percent of America's projected energy needs in 2010.

In addition to Sen. Wirth, several other members of Congress have sponsored legislation addressing global warming. So far, the bill introduced by Rep. Claudine Schneider (R-R.I.) has attracted the most bipartisan congressional support and the endorsement of many environmental groups, including the Sierra Club, the National Audubon Society, and the Natural Resources Defense Council. Environmental organizations have withheld their endorsements of Wirth's bill because Schneider's is more comprehensive and maps out a specific plan for reducing greenhouse gas emissions.

Asked to explain the lack of a nuclear component in her bill, Schneider responds, "Nuclear power is not cost-effective now, nor will it be in the future. My goal is to minimize cost, not just to throw money at an option."

Slowing the warming of our planet will be costly no matter what steps are taken. It will involve international cooperation to stop deforestation, increase tree planting, eliminate chlorofluorocarbon emissions, and reduce fossil-fuel combustion in transportation, industry, residences, and power plants. But taking no action at all would end up being even more costly and would mean a tripling of CO_2 emissions over the next 50 years. Investments in energy efficiency will not only reverse that, but will free capital for reforestation and other greenhouse-mitigating activities.

Nonetheless, nuclear power still has the upper hand over energy efficiency in at least two respects, according to David Hawkins, an attorney with the National Resources Defense Council. "We're dealing with human beings, whose psychology tends to be dominated by what appears to be the easiest fix. Somehow, we must focus on harder-to-grasp issues like energy efficiency. It's difficult, for we're sitting at a table with people who say, 'Here, we'll sell you the answer, in a technological box.'"

Hawkins points out that in addition to holding this psychological advantage, the nuclear industry tends to dominate the agenda of energy policymakers. "It's like going to a dinner party with Henry Kissinger," he says. "Everyone is going to pay attention to him, even if someone else has something better to say."

From Sierra, *March/April 1989. Reprinted by permission. Michael Philips is a project associate with the International Institute for Energy Conservation in Washington, D.C.*

Earth Care Action

City of Freeways Plants an Idea to Help Nature

By Robert Reinhold

If any city has a moral obligation to do something about global warming, this is it. And so the City of Angels and internal combustion took action.

Mayor Tom Bradley has announced a program to encourage the planting of two million to five million trees in the city in the next five years, trees that would serve to cool buildings and absorb the carbon dioxide that experts say is a major culprit in the worldwide warming trend.

Many scientists fear that the general increase in temperatures around the world, caused by gases absorbing solar heat, could bring environmental catastrophe in the next century.

Mayor Bradley said that although the Los Angeles tree program could have only a small effect on the warming trend, he hoped that major cities around the world would emulate his city and begin similar projects.

"We are trying to raise the level of consciousness, to spread information," Bradley said, just two hours after launching his campaign for a record fifth term as mayor of the second-largest American city.

He said he would write the mayors of Los Angeles's 14 "sister cities" abroad urging similar steps.

The mayor said the city would work with a local private group, Tree People, to plant the trees on parking lots, along streets, and on other unshaded places. He said the project would not cost the city any money beyond that for maintaining trees on city property, but he had no estimate of that expense.

"Environmental Crisis"

Joining Mayor Bradley for the announcement was Jeremy Rifkin, president of the Foundation on Economic Trends, a Washington-based public affairs group that has been sounding the alarm on the putative environmental hazards of genetic engineering and global warming. Rifkin spoke as a member of the recently formed Global Greenhouse Network, a group of organizations and lawmakers in 35 countries whose goal is to halve global carbon dioxide emissions by the year 2030.

Rifkin said he hoped that the tree program would be the first step in a worldwide effort to use local solutions to stem global warming without waiting for national and international bodies to act. He called such warming the "first truly global environmental crisis."

The warming is caused by various gases, particularly carbon dioxide produced by combustion of fossil fuels, chlorofluorocarbons, and halons from industry, methane from agriculture and waste dumps, and nitrous oxides from motor vehicles and fertilizers. These gases absorb radiation from the sun that would normally be reflected back into the outer atmosphere, and thus, they cause a gradual warming of the earth's surface. This effect is compounded by widespread deforestation for development and farming, which has reduced the number of green plants that absorb carbon dioxide.

There is debate among experts about how urgent the problem is and whether the warming has caused recent droughts and heat waves in parts of the world. There is a growing consensus, however, that concerted worldwide action is needed.

This month [January 1989] the National Academy of Sciences urged President-elect George Bush to put the issue high on his agenda. "Global environmental change may well be the most pressing international issue of the next century," it said.

Climatologists have predicted that the warming could melt the polar ice caps and thereby raise sea levels by several feet and inundate major cities. They also say the warming could drastically reduce the planet's capacity to produce food.

Rifkin said his organization approached Los Angeles first to tackle the problem on the local level because of its history of air pollution. California as a whole, he said, ranks with the top ten industrial countries of the world in contributing to global warming. With 23.5 million registered vehicles, California has about 13 percent of all vehicles in the United States; Los Angeles alone has about 5 million vehicles.

"Trees for the Olympics"

Today's tree-planting program is not the first in Los Angeles. A similar effort was made for aesthetic reasons before the Summer Olympic Games here in 1984, but many of the one million or so trees planted have since died for lack of proper care and planning.

John W. Stodder, Jr., the mayor's environmental aide, said that a better effort would be made to save the trees this time and that $25,000 had been raised privately for Tree People to study how to do that. Two other local environmental groups, Campaign California and the Coalition for Clean Air, are also involved in the project.

Mayor Bradley, who has long been associated with the economic growth that has contributed to the city's air pollution, asked the city council last week to adopt more than 50 measures to reduce air pollution. For example, he would eliminate most free parking in the city, even at shopping malls, use cleaner fuels at city utility plants, upgrade city vehicles to make them more fuel efficient, and create stronger incentives for van pooling and discouraging single-occupancy vehicles on the roads.

But the question remained today whether Los Angeles residents, wedded to an easy "car culture" more than any other Americans, were willing to make the compromises in their style and standard of life that may be necessary to clean the air.

From the New York Times, *January 12, 1989. Copyright © 1989 by the New York Times Company. Reprinted by permission.*

OZONE

Saving Our Skins

By Arjun Makhijani, Ph.D., Annie Makhajani, and Amanda Bickel

An 11-Point Plan to Save the Ozone Layer 46
How Orange Rinds Help Save the Ozone 49
Stepping Up to the Ozone Challenge 50
Vermont Moves to Save the Ozone Layer 52

Worldwide emissions of chlorofluorocarbons (CFCs) threaten to deplete upper atmospheric ozone—the very substance that shields us from harmful ultraviolet radiation. Under worst-case assumptions, the resulting increase in ultraviolet radiation could reach levels comparable to those following an all-out nuclear war.

For humans, the impacts of increased ultraviolet radiation (UV-B) would include severe, blistering sunburns after one to two hours of exposure to the sun. Outside work would become difficult or impossible. Cattle might have to graze after dark (if there were still any grass to eat). Entire ecosystems—such as coral reefs—might be wiped out. Even reducing our present-day CFC production by half might not avert this catastrophe.

The Montreal Protocol, signed in September 1987 by 24 countries, including the principal producer and consumer countries, requires a 50 percent reduction in CFCs by 1998. Unfortunately, the protocol's restrictions may not be adequate to protect us. Assuming that the Third World also cuts consumption by 50 percent from 1986 levels, the quantity of ozone-destroying chlorine in the atmosphere will still increase considerably.

From the Earth Island Journal, *Winter 1988/89. Reprinted by permission.*

However, the technologies to recover, recycle, and replace CFCs are available in most cases. If we act rapidly and thoroughly, and if international cooperation can be intensified, we believe that the production of CFCs can be phased out by 1995.

The Corporate Response

The plentiful availability of low-cost CFCs has led corporations throughout the world to use them carelessly. Many have failed to install recovery and recycling systems, even when such systems would be economical. The situation is somewhat reminiscent of energy use before the oil crisis of 1973, when many conservation measures were ignored.

Governments have left an enormous area of research and development in this large area to corporations, which, by and large, have pursued research only when the threat of regulation has been imminent. Indeed, after identifying some promising CFC substitutes during the 1970s, corporations abandoned the search for alternatives around 1980. They now cite lack of government controls as the reason for ending research. According to the Du Pont External Affairs Department, "In the absence of regulations there was nothing to drive the search for alternatives because there was no market demand."

This remarkable statement by the world's largest manufacturer of CFCs (25 percent of world production) has been echoed in the published literature of other corporations. It concedes that only governmental regulation will drive corporations to fulfill their environmental responsibilities, in part because the market by itself cannot do so. We want to add only that public vigilance and pressure are also critical to the process.

Du Pont enjoys sales of $30 billion per year, of which $600 million (about 2 percent) are CFC sales. Between 1974 and 1980, the company spent $2.5 million per year on research and development of alternative chemicals. Had corporate research and development continued, these chemicals would probably be available today, and a rapid and relatively easy phase-out of almost all major uses of CFCs would have been accomplished in three or four years.

We believe that the real reason that Du Pont and other corporations downgraded their research was the arrival of the Reagan administration. The Reagan-Bush ticket was elected on a platform promising to cut back drastically on environmental and other regulation of corporations. This impending change in the regulatory landscape appears to have persuaded Du Pont and others to abandon what Du Pont itself had described in June of 1980 as a "prudent" course of continued research.

Such evidence leaves us with an inescapable conclusion: The most important decisions regarding human survival and well-being cannot be left by default to corporations driven mainly by short-term profits.

The Technology Is Here!

In many areas, alternatives to CFC technology already exist. Some are even more economical. For example, the use of CFCs in cleaning electronic circuits can be replaced in most cases by cleaning with soap and water, a well-established technology. Digital Equipment Corporation has been using aqueous cleaning since 1974, and the U.S. Air Force allows its contractors to use it. (Ironically, U.S. Army and Navy regulations discourage its use.)

On environmental matters, corporations commonly insist upon having definitive evidence that their products harm human beings before they will consider change. One might view this as a sort of post-mortem approach to public health: Produce the corpse and then we will dissect it to identify the illness. Medical post-mortems at least help the subsequent victims of an illness. But when the corpse is the earth itself, there will be no subsequent patients.

As far as the ozone-depletion problem is concerned, the long term is already here. We must immediately phase out ozone-destroying materials from the world economy, we must find safe, economical alternatives and technologies, and we must institute these changes equitably around the world.

Two Million in the Bank

An important source of ozone-depleting chemicals are the stocks of chemicals already in the world economy. These break down into two categories. The first includes CFC uses that result in prompt releases into the atmosphere, either during manufacturing or upon consumer use of the product. Solvents, aerosol propellants, flexible foam manufacture, etc., fall into this broad category. The second covers applications such as refrigeration and closed-cell insulating foam; substantial portions of the already-produced CFCs are still stored ("banked") in these devices.

We estimate roughly 2 million tons of CFCs are banked in existing equipment, out of the total of about 16 million tons of CFC-11 and CFC-12 manufactured from the inception of CFC use in 1931 up through 1985.

Air-conditioning and refrigeration equipment contains a very large stock of unreleased

---

Earth Care Action

An 11-Point Plan to Save the Ozone Layer

By Arjun Makhijani, Ph.D., Annie Makhajani, and Amanda Bickel

1. Phase out currently regulated CFCs by 1995.

2. Phase out methyl chloroform and carbon tetrachloride by the year 2000.

3. Encourage recovery and recycling of ozone-destroying compounds.

4. Establish a tax on CFCs, chlorocarbons, and ozone-depleting substitutes.

5. Change U.S. Army and Navy regulations, which now encourage CFC uses in cleaning.

6. Enact an immediate, worldwide ban on CFC use in aerosols, except certain medical uses.

7. Implement coordinated studies by the United Nations Environment Program and the Environmental Protection Agency on the environmental impacts and toxicity of CFCs and alternative chemicals and processes.

8. Intensify research and international cooperation on ozone-depletion and global warming questions, on their ecological and economic implication, and on how these problems might be mitigated and ultimately resolved.

9. Implement a comprehensive plan to assist the Third World in phasing out CFCs and chlorocarbons, while adapting substitute chemicals and technologies.

10. Complete fulfillment and extension of the Montreal Protocol.

11. Implement consistent and firm regulation of corporations to assure environmental protection.

From the Earth Island Journal, *Winter 1988/89.*

CFCs. This "bank" of CFCs could contribute significantly to releases in the future unless appropriate actions are taken to collect and destroy them.

Worldwide, the use of CFCs for refrigeration is somewhat larger than in the U.S. If we assume that the proportion of CFCs stocked in existing automobiles, buildings, and appliances in the rest of the world is similar to the proportion stocked in the U.S., we can estimate the rest of the world's total CFC stock in refrigeration and air-conditioning at roughly 700 kilograms.

Thus, combining U.S. and world estimates, roughly one billion kilograms (one million tons) of CFCs are stored in existing refrigerators, chillers, air conditioners, etc.—about equal to the worldwide use of all regulated CFCs for all purposes in 1985.

Cool the Air; Heat the Planet

Mobile air-conditioning, principally in passenger vehicles, represents the largest specific end-use of CFCs in the U.S., and largely as a result, the third-largest use worldwide. CFC-12 is used as the coolant in mobile air conditioners, with one-third of it charging new mobile air conditioners and the other two-thirds used for detecting leaks and servicing and recharging existing systems.

Car air-conditioning is one of the most difficult and controversial areas for CFC phase-out. Car air conditioner capacity is primarily determined by the requirement that the air inside the car, as well as the solid parts such as the steering wheel, receive a blast of cool air moments after the car is started, often after having been parked in direct sunshine for a long period. In other words, almost all car air conditioners are designed for the conditions one might expect in midsummer Las Vegas or Phoenix. As a result, present car air conditioners are quite large—each has about two tons of cooling capacity (24,000 BTUs).

Car manufacturers claim that large air conditioner size is partly dictated by the need to operate in high-humidity areas. However, a 24,000 BTU air conditioner can keep a modest-sized house at a comfortable humidity level. In our opinion, the principal determinant of large air conditioner size appears to be "solar loading"—i.e., the need to dissipate large amounts of retained heat.

Even with such a large air conditioner size, the solid parts of the car cannot be cooled very fast since they store a considerable amount of heat. When the outside temperature is 110°F, the inside of a car with untinted glass parked in the sun without shade or ventilation will heat up to about 225°F—that is, exceeding the boiling point of water.

The large air conditioner size means that a relatively large amount of coolant must be used: about four pounds of CFC-12 per car. This also means that large amounts of CFCs are needed for leak detection and recharging. Therefore, reducing air conditioner size is important to reducing CFC automobile gasoline consumption.

Plugging the Largest Leaks

Two-thirds of CFC use for mobile air-conditioning occurs during the servicing and recharging of existing units. Refrigerant loss in a typical unit occurs within three to five years, necessitating service and recharging. The loss results because car air conditioners are not hermetically sealed units. Instead, they are engine-mounted—the most convenient and initially inexpensive design. Engine-mounting of the compressor subjects the system to considerable vibration, necessitating the use of flexible hoses and connectors. Refrigerant loss typically occurs as these flexible hoses and connectors loosen due to engine vibration. The flexible hoses are also to some extent permeable to the refrigerant.

A typical service routine for a leaky system involves the following steps: (1) venting the remaining refrigerant into the atmosphere; (2) after the remaining CFC-12 is exhausted, the system is flushed with CFC-11 or CFC-13, which is also discarded; (3) CFC-12 is reintroduced into the system, and a detector is used to locate leaks; (4) the system is repaired; (5) the system is evacuated and, for ten minutes, reloaded with CFC-12 vapor, then reevacuated; (6) the system is recharged, often using 14-ounce cans that cannot be fully emptied.

At each step at which CFCs are introduced, they are vented into the environment. The remaining CFCs in the cans are also discarded.

There are several simple, inexpensive strategies for reducing the CFC emissions from automobile air conditioners. These include:

- Better designs of hoses, gaskets, etc.

- Better diagnostic techniques for early detection of leaks.

- Recovery, reconditioning, and reuse of CFCs used during servicing and repair.

- Using alternative, non-ozone-depleting chemicals for essentially the same system design.

- Introducing totally new air-conditioning systems.

Recovery and Reuse

This area is critical to reduction of emissions in the short and medium term, even if CFC-12 in existing systems can be replaced in the long term by non-ozone-depleting "drop-in" substitutes. Recovery will be even more critical if no substitute becomes available in the next several years.

Recovery, reconditioning, and reuse of CFCs in cooling systems is already a well-established technology. However, it is generally applied to large, stationary systems (more than 50 tons). The reconditioning can be done on-site by the installation of in-line filters, or the refrigerant can be removed and purified off-site and stored for later use. One characteristic of the present recycling systems is that the reconditioned CFCs are held for reuse by the very customers with whom the CFCs originated.

Thus, the technology is already available for recovery of CFCs at every step at which they are vented during servicing and repair. Collection systems, on-site storage, and off-site reconditioning must still be organized. Service station personnel must also be trained to use the systems. All this will be a major undertaking.

In addition to servicing and repairing leaks, large quantities of CFCs are banked in existing auto air-coolers—more than 100 million kilograms in the U.S. and about twice that in the world. CFCs are generally allowed to escape into the environment when cars are sold for parts or scrap. Collection of CFCs at the point of automobile disposal will be a very important element in reducing emissions of CFCs from the existing banks of stored materials.

Alternative Chemicals

A considerable number of restrictions governs chemical substitutes for CFC-12. HCFC-22, which is one possible substitute, tends to diffuse through the present hose materials rather rapidly, so that either new hose types or system redesign would be required. Operating temperatures and gas pressures impose other restrictions. Any substantial change in these specifications requires system redesign.

The major automobile companies appear to have decided against any substantial commitment of resources to system redesign. Rather, they are counting on the availability of the most promising substitute to date—HFC-13a. This chemical is nonflammable and contains no chlorine, therefore, it is classified as having zero ozone-depleting potential. Du Pont claims that its greenhouse-effect level is less than 10 percent that of CFC-12.

While HFC-13a is a promising alternative chemical whose development should be vigorously pursued, we conclude that considerable risk attends the great reliance being placed on this one chemical. We urge an accompanying investment in alternative auto air-conditioning systems and system redesign. As noted, mobile air-conditioning contributed roughly five times as much CFCs to 1985 U.S. releases as all other air-conditioning and refrigeration applications combined.

Alternative Systems

System redesign can be approached in a number of ways, some complementary and others mutually exclusive. These include side vent windows at least as an option on cars. For vehicles in some areas this would provide a reasonable level of comfort, as it has in older model cars. The absence of side vent windows in cars reduces air flow into the car, making air-conditioning more desirable in climates where adequate ventilation would provide sufficient cooling for many people's tastes. This approach also has the merit of saving considerable amounts of energy. Up to 10 percent of the energy use in a car can be for air-conditioning.

Earth Care Action

How Orange Rinds Help Save the Ozone

By Peter Tonge

The throwaway part of your breakfast juice is helping to reduce a serious environment threat—the widening gap in the ozone shield over the Antarctic. A substance made from orange and grapefruit rinds is replacing the harmful chlorofluorocarbons (CFCs) in a new solvent.

William Galloway, of Fernandina Beach, Florida, says depletion of the protective ozone shield surrounding the earth by the use of CFCs "is a much more serious problem than most scientists are saying publicly." So six years ago, in collaboration with AT&T, he set his PetroFerm research company to tackle the problem.

Numerous environmentally benign products, including metal cleaners, an optical lens cleaner, and a substance used in asphalt extraction, have emerged from the company's research laboratories in recent years. Now comes BioAct EC7, a solvent based on terpenes that are extracted principally from citrus waste but also from pine bark and other organic materials.

BioAct EC7 can substitute for the CFC used as a cleaning agent in electronics manufacturing.

While the public is primarily aware of CFC as the now-banned (in several countries) propellant in aerosol cans, it remains a commonplace, highly evaporative cleanser, accounting for some 17 percent of its total use worldwide, according to Worldwatch Institute figures. Refrigeration/air conditioning (35 percent) and foam packaging and insulation (25 percent) are the leading users.

In particular, CFCs are used to wash printed wiring assemblies (PWAs) in radios, computers, guidance systems, and a host of other electronic gadgetry.

PWAs are the cards that computer chips are fastened to. They are circuit boards, or, in other words, "the parts that do the thinking," as Craig Hood, marketing manager for PetroFerm, describes them.

For this reason the boards must be spotless. A speck of dirt or flux in the wrong place could result in faulty electronic signals. "That might not be crucial in your transistor radio, but in the guidance system of a space probe, or on an aircraft, the end result could be disastrous," Hood points out.

When a circuit board is cleaned with a chlorofluorocarbon solvent, the fluid on the board is left to evaporate, and therein lies the danger to the environment. The solvent rises as a gas into the atmosphere, where it slowly degrades the ozone shield.

In contrast, BioAct EC7 is a solvent based on terpenes that occur naturally on a wide scale. Biodegradable, they do no harm to the environment. The oils pressed from citrus rinds left over from the juicing process are "particularly rich in terpenes," says Hood. This makes the product a natural for Florida.

AT&T has begun using EC7 because the product "outperforms CFC solvents and is already cost-competitive," Hood says. The news, he adds, "is all good."

Under international agreement, CFC production will steadily decrease over the next decade until, it is hoped, it will cease altogether. Meanwhile, some Florida citrus rinds are contributing to the process.

Reprinted by permission from the Christian Science Monitor, *February 7, 1989. Copyright © 1989 the Christian Science Publishing Company. All rights reserved.*

Earth Care Action

Stepping Up to the Ozone Challenge

By Julian Baum

Anxiety about deterioration of the earth's ozone layer is prompting unusual international cooperation to curb chemical pollution of the atmosphere.

Representatives of 124 nations heard pleas from scientists and government ministers at a London meeting this month [March 1988] to accelerate the phase-out of chemical substances that break down the thin layer of ozone protecting the earth from ultraviolet radiation.

The European Community (EC) set the pace. Its environment commissioner, Carlo Ripa di Meana, told the conference that the EC could eliminate the use of chemicals harmful to the ozone by 1996 or 1997. This is three to four years ahead of targets set last week, when the EC, in a surprise move, pledged to eliminate 85 percent of chlorofluorocarbon (CFC) usage as soon as possible and to impose a complete ban by the year 2000.

President George Bush has also announced that the United States is committed to a ban on the offending chemicals—mainly the family of compounds known as CFCs as well as halons—by the end of the century, pending the availability of safe substitutes that industry is pushing to develop. Canada and several Scandinavian countries have set similar goals.

"The world community has a strong sense of urgency about this problem, and we are now dealing with it with the seriousness it deserves," said William Reilly, head of the Environmental Protection Agency.

The U.S. is the world's largest producer and consumer of CFCs. Several U.S. senators have introduced legislation in Congress that would compel a phase-out well before the year 2000.

But experts say that even if the Western industrialized countries, which account for 70 percent of CFC use, do carry out such a ban, world use of CFCs will continue to grow unless the developing countries as well as Japan and the Soviet Union cooperate.

Ripa di Meana said that developing countries need financial and technical assistance to help them make the transition to ozone-friendly substances, and he proposed that Western Europe use its trade pact with African countries, the Lom´e Convention, as one mechanism of assistance.

With almost 40 percent of the world's population, China, India, and Brazil now account for about 2 percent of CFC use. Scientists are concerned that as these countries industrialize and improve their standard of living, their consumption of CFCs—widely used in aerosol sprays, refrigerators, and air conditioners and as a foaming agent in insulation—could overtake the amounts now used by developed countries.

British Prime Minister Margaret Thatcher told the meeting that the need to protect the ozone layer was not a case of some countries asking others to act. "It is a case of every country taking action if we are to protect all peoples," she said. She said that the London meeting demonstrated the urgency of the problem and that "no one can opt out."

Growing public awareness of the threat to the ozone layer has pushed many governments to affirm their support for a United Nations agreement, known as the Montreal Protocol, which pledges a 50 percent phase-out of CFCs by the year 2000. This week, 20 more countries announced their endorsement of the agreement, bringing the total number to 53, and over a dozen others have indicated they probably will sign.

However, most countries at the London meeting had not signed the protocol, and scientists now say that the agreement, drafted two years ago, is no longer adequate to prevent the rapid depletion of the ozone layer. Attention is now focused on a meeting in Helsinki in May to strengthen the protocol's provisions and to take the first steps toward its implementation. There will be a follow-up meeting in London in April 1990.

The rush to save the ozone layer is a reaction to the findings of research conducted primarily by British and American scientists. In 1987, a British Antarctic Survey station reported an "ozone hole" over the Antarctic, showing that 97 percent of the ozone was depleted over a six-week period during the seasonal peak of the ozone's formation. Another international survey this year showed that concentrations of ozone-destroying CFCs over the arctic region were 50 times greater than predicted, suggesting there could also be an "ozone hole" over the North Pole.

"The present pollution of the atmosphere is a fact we will have to live with for all of the 21st century," said Sherry Rowland, an atmospheric scientist with the University of California at Berkeley, who first sounded the warning about ozone depletion a decade ago.

Reprinted by permission from the Christian Science Monitor, *March 8, 1989. Copyright © 1989 the Christian Science Publishing Company. All rights reserved.*

Earth Care Action

Vermont Moves to Save the Ozone Layer

By Meg Dennison

Vermont is becoming the first state to ban car air conditioners that use a chemical linked to the destruction of the atmosphere's ozone layer, which protects the earth from ultraviolet radiation.

Car air conditioners are the largest source of the country's contribution to ozone depletion, according to scientists and testimony before legislative committees.

Gov. Madeleine M. Kunin has hailed the state's bill, which was passed May 2 [1989] and which she is expected to sign within the next two weeks, as "landmark legislation."

Starting with the 1993 model year, the legislation would ban the sale or registration of cars equipped with air conditioners that use chlorofluorocarbons, or CFCs, for coolant. Car air conditioners were singled out partly because they use CFC-12, considered the type of CFC most damaging to the ozone layer. Home refrigerators use a different type of CFC.

When CFC molecules rise in the atmosphere, they damage the ozone layer 15 to 20 miles up. Scientists say that lets in extra radiation, which increases the incidence of skin cancer and eye disease and damages plants and animals crucial to the food chain.

CFCs, developed as refrigerants in 1930 by a General Motors chemist, were considered ideal chemicals because they are nontoxic and noncombustible.

The chemicals are widely used as coolants and as propellants, such as in spray cans. CFC-propelled noisemakers and plastic party streamers, as well as noncommercial and non-industrial uses of CFCs for cleaning photographic or electronic equipment, would be banned next year.

Supporters of Vermont's legislation say it is the most far-reaching of any of the two dozen ozone-protection bills introduced in state legislatures this year.

Researchers at General Motors Corp., Ford Motor Co., Chrysler Corp., and their suppliers have been looking for ways to eliminate CFC-12 from their cars' air-conditioning systems.

Most of the research involves a substance called 134-A, which contains no chlorofluorocarbons.

But there are problems.

"There is no [compatible] lubricant, and the toxicological testing is not finished," said Norm Nielsen, supervisor of government regulatory programs at Chrysler. "Officially, it will take three or four years to finish."

From the Philadelphia Inquirer, *May 12, 1989. Reprinted by permission.*

PESTICIDES

A Mother's Crusade

By James C. McCullagh

Start Your Own Chapter 58
Big Firms Get High on Organic Farming 61
What's in a Name? 62
Indonesia Cuts Pesticide Use 64
What Really Happens When You Cut Chemicals? 66
Hometown Hero 69
Myth vs. Reality 70
Hormones to Be Used in Lieu of Conventional Pesticides 72
Please Don't Eat the Daisies 73
Lawn-Care Companies Now Offer Alternatives 74

Meryl Streep likes it to be Atlantic cool when she sleeps. She knew something was wrong when her Connecticut summer of 1988 was as hot as Houston in August. She knew that the world was going crazy, heating up like a sunburn. The seasons were supposed to be more predictable than her movie roles. The global warming had penetrated her family sanctuary in the Berkshires—a long way from the urbanites' view of "civilization"—and she wasn't going to take it anymore. So she did what any other red-blooded American mother who happens to be the country's best actress would do: She called Robert Redford.

She called him for her children's sake. A long-time environmentalist by politics and inclination, Streep had supported "various causes with money, hoping other people would do the work." But those 100-degree summer nights convinced her that she had to do something. "The events of the past summer pressed environmental concerns on all of us." Enter Redford, who has "done this sort of thing for years." That telephone call during Streep's hothouse days started a chain of events that will likely have a profound effect on the food we feed our children in the future.

From Organic Gardening, *April 1989. Reprinted by permission.*

PESTICIDES

Meryl Streep and other concerned parents take action for pesticide-free food at a meeting in Falls Village, Connecticut. (George DeVault/New Farm)

Famous for her roles in *Out of Africa*, *Silkwood*, and *Sophie's Choice*, Streep comes to this drama as the mother of three, a woman who moves through roles like quicksilver. If, as one writer has suggested, Streep's essence in her movies is "control—the unwillingness to reveal or surrender the self," her essence as a mother is to "go absolutely crazy" when she thinks her children are being harmed. And she thinks that they are.

In a kind of serendipitous intersect, Robert Redford sent Streep to the Natural Resources Defense Council (NRDC), a well-respected nonprofit organization in Washington, D.C., dedicated to improving the environment. Redford said the NRDC was one of the "most effective, respected environmental groups in the country." Furthermore, it would respect her star quality. At the time, the NRDC just happened to be completing a final draft of a blockbuster report, *Intolerable Risk: Pesticides in Children's Foods*. The report focused Streep's growing environmental concerns on this issue, which affects parents everywhere.

That summer of Streep's general discontent had led her to a spring issue that she could embrace as mother and actress. A woman who shuns interviews, she now has begun a media blitz to spread the word about the damage pesticides are doing to our children. If there's ever a script she's taken to heart, this is the one. No wonder.

After an exhaustive two-year study of available data, the NRDC concluded that our nation's children are being harmed by the very fruits and vegetables we tell them will make them grow up healthy and strong. These staples of children's diets routinely—and lawfully—contain dangerous amounts of pesticides, which pose increased risk of cancer, neurobehavioral damage, and other health problems.

According to the NRDC, federal pesticide residue standards are woefully inadequate because they are based on an average adult's diet and do not take into account that children receive greater exposure to pesticides than adults due to their much greater consumption of produce. In addition, existing standards may not reflect the young's increased vulnerability to toxic chemicals.

The findings of this report are startling. The average preschool child consumes 6 times as much fruit as the average adult, and 18 times as much apple juice—and the average toddler consumes 31 times as much applesauce.

The NRDC states that the average preschooler receives 400 percent more total exposure than an adult to the eight carcinogenic pesticides analyzed. As a result of this massive, poorly regulated exposure, many children are being exposed to pesticides that can cause cancer and harm the brain and nervous system.

Children are at particular risk because the Environmental Protection Agency (EPA) has historically ignored infant and child consumption patterns when it set the current legal limits for almost any pesticide in food. Instead, the agency based its legal limits (known as tolerances) almost entirely on estimated adult consumptions (which are probably too low, but that's another story). For example, when setting most limits for pesticide residues in food, the EPA underestimated preschoolers' consumption of grapes by 600 percent and apples and oranges by 500 percent, according to the NRDC.

Furthermore, government regulations may inadequately account for children's increased susceptibility to carcinogens. The NRDC report states that the young are frequently more vulnerable to these toxic substances than adults. This is because children can retain a greater portion of a given dose (because, for example, the gastrointestinal tract is more permeable), and also are not as capable of keeping toxins away from organs in the body or excreting toxins out of their bodies.

What is equally alarming about this report is that the NRDC was very conservative in its claims. For example, it selected only 23 of the 300 pesticides legally used on foods, including daminozide, mancozeb, captan, methamidophos, parathion, methyl parathion, and diazinon. It chose this sampling because they are frequently used and because information exists on their levels in food.

Many commonly used pesticides known to be hazardous were not included in the study because information on their residues in food was lacking. The study also only analyzed pesticides found in 27 fruits and vegetables. Corn, grain, and milk products were excluded. The NRDC notes that the extent of actual pesticide contamination of the nation's dairy products and meat supply remains relatively unknown because testing methods cannot detect many of the residues, particularly of the metabolites.

The Thoughtful Flight

Sitting in a nondescript meeting room in New York City, Meryl Streep lists the major findings of the NRDC as if she is the principal author. If she isn't, she's the principal owner, rolling the numbers around in her heart, cutting through the data like a knife. "It really got to me. Food is a religious, sacred thing. The pesticide report turns the world inside out. To think that food could hurt a child. It's weird. It upsets every corner of my existence. I went sort of wild. It's insane that we have to worry about the food we give to our children.

"With children, you don't take any chances," she says bluntly, with a determination that doesn't particularly need the support of research. "We don't offer children cigarettes and alcohol. Why should we offer them food contaminated with pesticides at dangerously high levels, even when we don't know all the health risks involved?"

These words are from a mother to mothers across the country. Streep readily admits that before she had children she would have treated the NRDC report as a second-page news item. Now it's personal. The food she brings into her home is contaminated. "It's different now," she explains. "Tonight I go home and make dinner for a two-year-old." She's incredulous. "Every single thing that goes into their mouths has been sprayed."

Streep is full of passion and rage. She has no patience with the EPA, which, she is convinced—on good authority—sat on information that would have brought this matter to the public's attention years ago. She has no time for bureaucracy that doesn't protect the "little ones; the most vulnerable members of our society."

In many ways, the script is her own, a fact that might make the NRDC's lawyers squirm. She wonders out loud whether increased pesticide use in the last decade has anything to do with the increased incidence of infertility in women.

Her intuition tells her that "fetal tissue must be more vulnerable." She is very worried about the damage these poisons are doing to children while the government slouches toward some solution. She is worried about what we don't know, a fear echoed by Herbert Needleman, M.D., professor of child psychiatry and associate professor of pediatrics at the University of Pittsburgh. "We are now in the area of pesticides that we were with lead 30 years ago. We shouldn't have to repeat history to safeguard our children. The toxicological data demonstrate that developing brains are more sensitive that adult brains.

"Children have extraordinary body burdens. The younger brain is more susceptible to insults than fully developed ones. The government should revise the tolerance thresholds for pesticide residues. This is vital, and unnecessary use of pesticides should be reduced to the vanishing point," he proclaims.

The NRDC report fully endorses this view. So does Meryl Streep. But she's not taking any chances. She knows about the dwarf speed of government. Besides, she knows that if you want to turn the food industry on its ear, you have to go to the people who make a difference. First stop: other women. Right in her own backyard.

From the Ground Up

This gifted actress joined with Wendy Rockefeller to organize the first local chapter of Mothers and Others for Pesticide Limits, which is modeled on Mothers against Drunk Driving and is expected to grow like garden clubs across the country. According to Streep, this group is not just a bunch of yuppies, ex-politicians, and one movie star. "What we have is 20 women in a living room who are not really joiners. They are holding things together. The uniform reaction was, 'We've got a problem. Let's solve it.' "

Streep makes no bones about the reasons for her involvement. "I had no choice but to get out in front on this issue," she says. "My fame gave me a unique opportunity. It's in my family's interest to have safe food in Connecticut."

The next stop was the local produce manager, to convince him that it's in the store's best interest to carry organic, pesticide-free produce. To Streep, "It's bizarre that the produce manager is more important to my child's health than the pediatrician."

This woman is a strategist, a tactician. She knows that "women will be passionate about the issue in that it affects our most vulnerable group. People will relate to that." On the other hand, she knows that her organization will have to convince produce managers to carry pesticide-free fruits and vegetables. She recommends going to local stores and "demanding clean produce."

The fledgling Connecticut group is raising funds, preparing buyers' guides, campaigning in schools, and working with local farmers. She reports that in the first month "three local farmers already are interested in working with us." Streep is concerned that any "new regulations not burden American farmers," and she would like to link the organizations with farmers who have a commitment to sustainable or organic methods.

Although Streep thinks the best hope for this issue is for mothers across America to organize chapters of Mothers and Others (see "Start Your Own Chapter" below), as a tactician she will be seeking the support of powerful women across the country, including senators and business executives. She plans to address a national conference of the PTA.

Earth Care Action

Start Your Own Chapter
By Christopher Shirley

You can organize a local chapter of Mothers and Others for Pesticide Limits in your neighborhood. To start, follow these tips, based on the experience of Roberta Willis, a founding member of the original chapter in Connecticut.

First, recruit enough members to help with activities and information gathering. "We started with friends and word-of-mouth, and quickly grew to a core group of 25 to 30, mostly mothers," says Willis. Next, form committees based on member interests and group goals. "Our aims were self-education and increased community awareness." Her group formed three committees: local food sources, education, and organic farming, the latter to promote local alternatives to large-scale agriculture.

"Our food sources committee researched and listed local organic farmstands and producers, mail-order organic suppliers, and organic garden supply companies," says Willis. "We also met with supermarket officials to express concern over pesticide residues and to promote the need for organic produce sections."

Next, establish affordable membership dues to fund office expenses and help with publicity. Write your own press releases and print fliers to promote your activities and to encourage membership. Seek free legal advice and, depending on state regulations, file incorporation papers to allow fund raising.

As a resource guide, Willis recommends *Pesticide Alert,* by Laurie Mott and Karen Snyder (Sierra Club Books, 730 Polk Street, San Francisco, CA 94109, $9.95 ppd.) For further information and for a copy of *For Our Kids' Sake: How to Protect Your Children from Pesticides in Food,* send $7.95 to Mothers and Others for Pesticide Limits/NRDC, P.O. Box 96641, Washington, DC 20090. For additional resource material, the following organizations can help: Americans for Safe Food, 1501 16th Street NW, Washington, DC 20036; Pesticides Action Network, Box 610, San Francisco, CA 94101; Northwest Coalition for Alternatives to Pesticides, Box 1393, Eugene, OR 97440; and the National Coalition against the Misuse of Pesticides, 530 7th Street SE, Washington, DC 20003.

From Organic Gardening, *April 1989.*

Most intriguing to Streep is that she sees a sea of change in national consciousness, a tide running far ahead of the stagnant waters in Washington. She recalls that during the recent presidential election, voters voiced a keen interest in environmental matters. Her intuition is supported by an *Organic Gardening*/Harris poll (in the June 1989 issue) indicating that American consumers are overwhelmingly in favor of organic produce. If they had a choice, more than 70 percent of Americans said they would buy organic fruits and vegetables at their local market. More than 50 percent would pay up to 20 percent more for them. Those who have eaten organic produce say the most important reason for doing so is long-term health.

As usual, the Food and Drug Administration (FDA), which is responsible for labeling foods, is completely out of step with the American consumer. In December, the FDA repeated its long-standing position: "We believe it would serve no useful purpose to create standards for foods that are virtually the same regardless of method of production," said John M. Taylor, associate commissioner for regulatory affairs.

Streep thinks that if Americans are given the information, they will opt for "clean" food. Consider, for example, the case of Keebler Inc. Pressured by a consumer advertising campaign, the packaged-goods snack manufacturer stopped using highly saturated coconut and palm oil in processed food. And General Mills, Ralston Purina, Borden, Pillsbury, and Quaker Oats have followed suit. She also is heartened by the organic certification programs of states such as Texas and California and strongly supports labeling. "Yes, I'd like to see code structures federally mandated and stronger regulations." She lofts a polite shot across the political bow. "I hope the people in D.C. realize that people will demand that this enforcement be built into the system."

Lest anyone think that Streep brings a particular viewpoint to this table, she is quick to point out that she is not a vegetarian. "I eat a lot of vegetables and grains, but I love beef, pork, and lamb." However, she sees Mothers and Others expanding in time to include scrutiny of meat and grains. Anything that "intrudes into the relationship between mother and child" is fair game.

Streep and the NRDC are of one voice in their recommendations. The EPA should immediately revise its pesticide limits to ensure that children are protected. The FDA must improve its ability to detect the residues remaining on food and stop the sale of food that contains dangerous levels of pesticide residues. The U.S. Department of Agriculture (USDA) must provide growers with incentives to minimize pesticide use and to use farming methods that do not rely on dangerous pesticides.

In the short run, consumers can carefully wash and peel produce, eat domestically grown fruits and vegetables that are in season, and seek out food grown by organic gardeners and farmers. And of course, they can garden.

Streep, a weekend vegetable gardener whose real passion is wildflowers and reading seed catalogs, is excited about the possibilities for gardening. The message she will take to the national PTA is that every school should teach gardening. Much as she wants the federal agencies to move quickly, she has far more confidence in the American people. To her, there's a "commonsense ingredient to the report that will appeal to the country."

But she's a realist. "I can't do battle with the biochemists," she acknowledges, "but I don't think I'll have to. Though the NRDC report data is solid, mothers will respond to this on an emotional level. When kids are involved, it's a completely different set of circumstances."

With no attempt at hyperbole she says, "I want this movement to be huge, huge. Community gardens everywhere."

A People's Revolution

Streep slips into the fifth row of the auditorium in Housatonic Valley Regional High School, Falls Village, Connecticut. It's Friday night and the place is packed. Camera crews from "60 Minutes" have the space abuzz. No one seems to pay a whole lot of attention to Streep. She likes it that way.

This is the first public presentation of the NRDC findings. The forum is a public symposium hosted by the Northwest Connecticut Chapter of Mothers and Others. In a few minutes the cameras are forgotten. These people are serious; they know something about leverage and political action. The man in front of me whispers to his wife, "This is the way to change society."

I can't help but remember that this is Inauguration night. Washington is alive with parties and socials celebrating a campaign that said little about our country's future, even less about the environment. While President Bush and his entourage sway to the music, the folks in northwest Connecticut are taking up democratic arms to protect their children and the nation's children. For three hours the country seems to be in good hands.

It's one thing to read the cold data of the NRDC report, entirely another to hear the data presented with concern and emotion in an auditorium full of mothers and fathers who are visibly shaken by the news. Robin Whyatt, NRDC staff scientist and author of the recently released NRDC report, armed to the ceiling with charts, speaks as a mother about this "major public health problem" that exposes the majority of our children to cancer risks.

Dr. Needleman, whose work with lead poisoning and his intuition tell him that pesticides are a real threat to children, wonders "if we want to wait 40 years" during which time "millions of our children will pay the price of avoidable exposure."

Through all of this, Streep is intense, nodding in agreement, chatting excitedly with the friends who surround her. She beams when Steven Markowitz, M.D., of the Division of Environmental and Occupational Medicine at Mount Sinai School of Medicine, asks, "Do we aggressively protect our children or wait for the ironclad proof that will never come?"

There is no question where she or the audience stands. Dr. Markowitz has a "strong hunch" that the NRDC's findings are right on target. The overflow audience is appreciative of these medical scientists who are ruled by common sense and intuition, as well as numbers.

Janet Hathaway, NRDC staff attorney, catches the spirit of the issue and occasion, giving the most powerful and moving presentation of the evening. "We are relying on the government to give us safe food," she says, "and they give us platitudes." She is unequivocal in her advice. "Consumers have to demand safe foods. I shouldn't have to be an analytical chemist to eat a safe breakfast. We've got to say we've had enough. We've got to change how we grow food. Legal food must be safe food."

Streep's main question of the night deals with the substance daminozide/UDMH (trade name Alar), which is found in rocket fuel and also sprayed on apples to extend shelf life, improve appearance, and expedite harvest. Daminozide breaks down into UDMH, its metabolite, during the processing of apples for applesauce. She wants to know how the USDA expects us to wash away that substance.

These people mean business. A fruit grower chides consumers for demanding cosmetically beautiful produce at their supermarkets—which can usually only be achieved through the use of pesticides. A woman complains that an organic orange looks like the "inside of a muffler" but once peeled, the fruit is simply beautiful. Someone suggests the Surgeon General should bring the weight of his office to bear on this problem. There also is talk of a national version of California Proposition 65, a law that requires the Governor to publicly list all known carcinogens.

This is the beginning of the revolution Streep is talking about. Already in her corner of Connecticut, stores are selling pesticide-free produce. Consumer demand is sure to spread, especially on the heels of her media blitz.

When Streep disappears into the night, scores of citizens remain to strategize. That is to her liking. But as much as she wants this to be a grassroots uprising, no one should expect her to leave this life-threatening issue soon. After all, our children are still at risk.

Earth Care Action

Big Firms Get High on Organic Farming

By Sonia L. Nazario

Organic growing is going mainstream, but that isn't cause for total joy down on the farm.

Even before the recent scare over tainted fruit from Chile and the controversy over apples sprayed with Alar, organic farming had been storming some unlikely bastions.

A month ago, for example, Jack Pandol, Sr., stood up at an agribusiness meeting in politically conservative Kern County, California, and pledged to aim to convert all of his 5,000 acres of produce to organic methods by 1995. Mouths dropped. Fellow pesticide-using farmers were dumbstruck.

Pandol, after all, is a leading county Republican. Here he was, announcing that he was diving headfirst into what has traditionally been a counterculture business run by flaky hippies.

Meeting Demand

Pandol isn't alone. In the past few months, several large Kern County growers have forsaken synthetic pesticides and fertilizers on more than a thousand acres. Nationwide, such big suppliers as Sunkist Growers Inc. and Castle and Cooke Inc.'s Dole Foods Co. subsidiary are starting to grow organic produce. California Certified Organic Farmers, or CCOF, a trade group, estimates that sales of organic produce and products in the U.S. doubled to $1 billion last year from 1983.

"It's a clear shift, almost a rush now," says Robert Scowcroft, executive director of CCOF, which has been deluged in recent weeks with calls from farmers wanting to switch. Betty Kananen, director of a Bellefontaine, Ohio, organic trade group, says, "The phone calls have tripled lately. I've about run my legs off. I was glad to go out and milk the cows."

For consumers, the result may be lower prices. But some organic growers fear that newcomers may lower standards—and may drive longtime producers out of business.

The shift to organic methods has been speeded by recent events. Kings Super Markets Inc. has received thousands of calls and letters in recent weeks from worried consumers seeking organic produce. Sales at Mrs. Gooch's, a Los Angeles natural-food chain store, have soared 20 percent in three weeks. "Are the chickens safe? Are the eggs safe?" an agitated customer demands at one Mrs. Gooch's store.

But even before the recent apple and tampering scares, chemical-using farmers were converting in growing numbers to meet demand. In a Louis Harris poll conducted for *Organic Gardening* magazine, 84.2 percent of Americans said they would buy organically grown food if available, and more than 50 percent would pay more for it. So, in the past six months, some 20 big supermarket chains have started stocking organic produce, according to Organic Farms Inc., the nation's largest organic distributor, whose sales more than doubled to $22 million in 1988. Some restaurateurs have switched to organic food because of what they say is its superior taste. Demand far exceeds supply, and merchants are scrambling.

For the hippies who started the organic movement in the Age of Aquarius, all is not peace and harmony. They are pleased that their beliefs and methods are being vindicated, and they realize that only these new, larger players can supply the increasing appetite for organic foods.

But they also fear for their own survival. Many have gotten by only because organic produce commands premiums of 15 to 30 percent. As more farmers enter the field, prices may plunge. If they are squeezed out, longtime organic growers say, the back-to-the-land dream of small-scale farming might also be shattered.

"Any of the big companies can come in and

What's in a Name?
By Sonia L. Nazario

Markets increasingly are applying labels to fruits and vegetables to indicate decreased use or absence of synthetically compounded chemicals and fertilizers. The following are some of the most frequently used terms and what they mean.

No-spray. Pesticides and other chemicals aren't sprayed onto the crop, but synthetically compounded fertilizers may be used.

Sustainable. May have some pesticides and other chemicals applied, but attempts are made to use reduced levels in conjunction with biological control methods to manage rather than kill pests.

Transitional. Crops grown on land prepared with organic techniques, but not long enough to be certified as organic by a state or trade group.

Organic. Synthetically compounded chemicals and fertilizers aren't used, although some naturally occurring chemicals may be employed on crops.

Biodynamic. The farmer plants and cultivates crops "in harmony" with the earth's seasons and utilizes organic materials native to the region.

From the Wall Street Journal, *March 21, 1989.*

blow me out of the water," says William Denevan, who started farming organically in 1974, after what he describes as a life of avoiding the military draft, distributing health-food products, surfing, and living in a tepee. "It scares the hell out of me," adds Denevan, whose 230-acre Happy Valley Farm in hills outside Santa Cruz, California, has been so swamped with calls for organic apples that he's taken the phone off the hook.

Prices already are falling as larger, more efficient competitors make inroads, some organic farmers say. Sy Weisman, who farms three acres of organic produce in Sebastapol, California, notes that the price of his radicchio has plunged to $3 a pound from $3.50 last year. Weisman has switched to other specialty crops, such as Persian basil and epazote, a Mexican herb popular in bean dishes.

Some organic farmers already decry the methods of a few large organic players. For example, Stephen Pavich, who has 2,700 organic acres in California and Arizona, injects some of his land with methyl bromide, a potent sterilizing fumigant, before planting new grapevines. Pavich waits longer than required by CCOF and state regulators before harvesting organic grapes for sale, but the old guard says such practices aren't in keeping with the spirit of organic farming, which strives to increase organic matter in soil and create sound ecosystems. Pavich says there aren't any natural alternatives to kill nematodes.

Most farmers continue to embrace the belief that the U.S. can't increase yields, lower food prices, and feed the world without using chemicals. But a number of case studies have found that yields and profits can be just as high on an organic farm as on a nonorganic farm. Organic ventures are more labor-intensive, but pesticide prices have been soaring, evening things out. John Givens, who farms 40 acres organically in Goleta, California, last year netted $45,000 and drives a baby blue BMW.

In recent years, organic farmers have had at their disposal bacteria that make worms explode, natural soaps that can be used as pesticides, and a range of beneficial pests that will prey on bugs harmful to their crops. Generally speaking, organic farmers shun synthetic chemicals and fertilizers and rely on crop rotation, manure, and biological methods to control pests and increase yields.

As the organic business grows, its lack of regulation becomes apparent. Definitions of "organic" vary widely, and only 14 states have special organic-food regulations. In most states, the task is left to myriad trade groups, raising potential conflicts of interest.

Organic Almonds?

In California, many organic farmers have been decertified by a trade group for falsely billing their products as organic. But when a California grower last year falsely claimed his almonds were organic and was decertified by CCOF, the farmer simply sought and got certification from another trade group, the trade group says.

Despite these hurdles, CCOF's Scowcroft predicts that in a decade "most regions of the country will have certified organic produce readily available most of the year." A report due this summer [1989] from the National Academy of Science's agriculture board is expected to conclude that while organic farming is economically viable, federal farm programs have set up strong disincentives to its practice. Still, although less than 1 percent of the nation's produce is now organic, and only some 30,000 of the country's 2.2 million farms grow organically, most organic farmers expect the percentage to increase to 8 or 9 percent in the next few years.

"In the future, people will no longer have individual control over their polluted air or water," says organic grower Paul Kolling, standing in his Sebastapol apple orchard. "The only thing they'll really be able to control on a daily basis will be the food they put into their mouths."

Reprinted by permission of the Wall Street Journal, *March 21, 1989. Copyright © 1989, Dow Jones & Company, Inc. All rights reserved worldwide.*

Earth Care Action

Indonesia Cuts Pesticide Use

By Kay Treakle and John Sacko

After years of relying on chemical insecticides to control rice pests, the Indonesian government has adopted integrated pest management (IPM) as a national policy and banned the use of 57 pesticides on its important rice crop. As a result, Indonesian farmers have dramatically cut their use of toxic pesticides and increased yields in rice paddies across the archipelago.

Indonesia is the first country in the world to take the radical step of institutionalizing a pest-control strategy by presidential decree. The commitment of President Raden Suharto to IPM reverses two decades of agricultural practices brought on by the "Green Revolution," which promised miracle rice yields but ultimately delivered devastating crop losses and an expensive chemical addiction.

Indonesia got started on the pesticide treadmill in 1967 when, upon taking office, President Suharto pledged to make his country, then one of the world's largest rice importers, self-sufficient in rice. The ingredients for his green revolution were simple: grow strains of rice specially developed to stand up straight, produce more kernels, and resist insects and diseases. Add irrigation and heavy doses of pesticides and fertilizers, and the technological package is complete. By the early 1980s, Indonesia's impressive yields made this country of 170 million people self-sufficient in rice.

But the economic and ecological cost of self-sufficiency was high. Government bureaucracies, under the influence of high-tech agricultural research institutions and powerful chemical companies, pushed a calendar-based chemical application as the solution to all the farmers' pest problems. Government loan packages invariably required that if a farmer accepted a loan, he or she had to accept the chemicals. The pesticide treadmill was virtually guaranteed by some of the highest pesticide subsidies in the world. During the first decades of Indonesia's Green Revolution, the government subsidized as much as 85 percent of pesticide costs, spending an astonishing $150 million a year.

The widespread adoption of the chemical approach left many farmers with the idea that pesticides are progressive and modern. With the cost of chemicals inconsequential, the prevailing attitude became, "the more the better."

But the "progressive" pest-control strategy

eventually wreaked ecological havoc on Indonesia's rice fields and turned Indonesia's green revolution brown. During the 1970s, a small flying insect called the rice brown planthopper destroyed millions of tons of Indonesian rice. Using typical Green Revolution methods—the introduction of a new strain of rice, IR36, and an arsenal of chemicals—the pest was suppressed. But only temporarily.

In 1986, the planthoppers were back. And by the end of the season, 10,000 hectares, roughly 600,000 tons, of rice were devastated. Chemical applications, in some cases up to 15 per month, did nothing to prevent disaster.

The fact is, for Indonesia's rice fields, along with those in Thailand and the Philippines, the use of pesticides actually contributed to the destruction of rice by the planthopper. Long-term studies now show that there is virtually no correlation between the use of pesticides and increased rice yields.

The conclusive evidence was developed by scientists with the Gadjah Mada University and the Food and Agriculture Organization (FAO) of the United Nations, along with Dr. Peter Kenmore, then a graduate student from the University of California at Berkeley. Their work concluded that the use of pesticides killed off the planthopper's natural enemies, including wolf spiders, lady beetles, and wasps. The pest problem was further exacerbated by spraying such highly toxic pesticides as methyl parathion and diazinon, which actually stimulated the reproduction of the brown planthopper. An FAO report concluded that the planthopper "was a pest because of, not in spite of, pesticide applications."

Having discovered the cause of the hopper infestation, the problem then became how to promote its cure. Neither President Suharto in Jakarta nor the rice farmer in the fields would have willingly given up pesticides without an effective alternative. Fortunately, that alternative, IPM, was then being successfully tested in the field.

The IPM approach involves conserving the hopper's natural enemies, manipulating the planting time to interfere with the steady supply of planthopper food (rice), and spraying only when the pests actually became a threat. At that point, farmers would apply chemicals designed to do the job without stimulating planthopper reproduction.

Translating IPM from field tests to a national policy was a monumental undertaking, which has become one of the most progressive and successful agricultural extension programs in the world. Based on a bottom-up approach to decision making and training developed by an interdisciplinary team of scientists, including anthropologists, entomologists, and social scientists, the Agricultural Extension Service was able to train more than 15,000 farmers in the IPM techniques in the first year. The decentralized approach worked like a telephone tree: 30 trainers taught IPM to 200 master field trainers, who in turn trained farmers in groups of 20 to 30. These groups subsequently became the nucleus for training other farmers. The government also supplied 1,000 motorcycles so that trainers and pest observers could swarm around the countryside monitoring pest infestations and teaching farmers how to cope without pesticides.

During the first season, pesticide use dropped drastically, and in the third season since the IPM decree, overall insecticide use fell 90 percent. During that same period, rice yields were up, from 6.1 tons per hectare in 1986 to 7.4 tons per hectare in 1987. Farmers' costs dropped from 7,500 rupiah in 1986 to 2,200 rupiah in 1987, despite an overall increase in the price of the chemicals.

Since Suharto's decision, more and more farmers have laid down their backpack sprayers and are taking greater responsibility for managing their crops. Increased incomes should help the economy, and the government will be able to use money previously used to purchase pesticides to further implement IPM.

The only losers in the IPM deal are the pesticide producers, including such multinational giants as ICI, Ciba-Geigy, Bayer, Monsanto, and Hoechst. Dow Chemical Pacific lost 80 percent of its business almost overnight, including half its shipment of Dursban, which was ordered before the ban. Dow has also had to postpone construction of a $4 million Dursban production facility. And Ciba-Geigy is now reconsidering plans to build a $10 million organophosphate manufacturing plant.

Petitions against the order have been submitted to the government, and the companies have complained that the severity of the brown planthopper threat has been exaggerated by the Indonesian government in order to justify a switch in agricultural policy.

But Indonesia has no intention of bowing to industry pressure. The government plans to train 2.5 million farmers by 1994. And other Asian countries—including such important rice producers as China, the Philippines, Malaysia, Thailand, and India—have recently developed similar projects with FAO assistance.

From Greenpeace *magazine, March/April 1989. Reprinted by permission.*

Earth Care Action

What Really Happens When You Cut Chemicals?

By Mike Brusko

Skeptical farmers have a standard litany of worries they repeat every time someone asks the question, "What's keeping you from cutting back on chemicals?"

Yields and income will fall. Pests will get out of control. Labor and equipment costs will increase. And perhaps worst of all, commodity bases and government payments will decrease.

But if you take the discussion out of the coffee shops and into the fields of experienced low-input farmers, you'll hear a much different story. "Conventional wisdom on reduction of inputs is not what it would seem to be," says Kevin Carroll, a graduate researcher in the University of Wisconsin's (UW) College of Agriculture and Life Sciences. "You may not have problems in areas you think you will."

Chances are yields and income won't fall, and pest pressure won't be intolerable, he says. Then again, some problems may pose more hardship than expected. Finding reliable advice on alternatives to chemicals, for example. Or handling criticism from neighbors.

Carroll and fellow UW researcher Margaret Krome last year [1988] set out to separate these myths and realities of low-input farming. With partial funding from the Wisconsin Department of Agriculture, Trade and Consumer Protection, they conducted surveys of two groups of Wisconsin farmers.

Krome, director of a sustainable agriculture program for the nonprofit Wisconsin Rural Development Center, chose a sample of 389 conventional farmers using all types of farming methods. Her goal was to find out what potential or *perceived* obstacles keep them from using fewer chemicals. "What do they say are the problems, and where does that lead us in terms of policy reforms that are needed?" she asked.

Carroll surveyed 177 farmers, all of whom had reduced or eliminated chemicals or were in the process of doing so. "My purpose was to look at what kinds of problems farmers face when they *actually* cut back on chemicals," he says. For those who had cut back, he adds, "It took about four years to complete the transition."

When the results are analyzed, Wisconsin agriculture officials will use them to help plan and monitor their new Sustainable Agriculture Demonstration Program. But the preliminary findings already reveal some useful trends, the researchers say.

Perception

Not surprisingly, fear of lost yields and/or income is the biggest perceived barrier to low-input farming among the conventional farmers surveyed. "They thought that by reducing chemicals, their profits would drop," says Krome. More than two-thirds of her respondents expressed that worry, and almost the same number listed it among their top three reasons for not trying low-input methods.

Another fear is that weed and pest problems are too severe to control with fewer chemicals. Some 38 percent of Krome's respondents worry about this.

There's also concern that cutting chemicals will mean buying a lot of new equipment. More than 45 percent of Krome's respondents think that would be the case, and more than 35 percent listed it among their top three reasons for not cutting chemicals.

Finally, says Krome, "It appears there are a lot of farmers who don't think they have an environmental problem." More than half the farmers in her survey said their chemical use doesn't harm the environment or family and livestock health. About three-fourths said they've already cut chemicals as much as they can, and two-thirds listed that among the top three reasons they don't reduce chemicals.

Reality

"Yield losses didn't seem to be a big problem," says Carroll, who surveyed the low-input group. Only 35 percent of his respondents said yields fell when they cut chemicals, and just 33 percent said yield losses actually cut profits. "More of the low-chemical farmers (76 percent) showed an increase in profits from savings on inputs than loss in profits because of loss in yields." That was true of all farmers in his survey, although cash-grain producers were a bit more likely to report yield losses then were dairy farmers, he notes.

Carroll suggests these results could vary in a state with fewer dairy farms and less land in pasture. "Wisconsin is tailor-made for sustainable agriculture," he says. Plus, factors like weather, commodity prices, and current debt load weren't included in his survey, making it hard for farmers to isolate the role cost cutting played in their incomes. "That's the kind of thing you can only answer on a case-by-case basis," says Carroll. "But I think you can say these farmers didn't have any big swings in profit either way."

The same goes for weed and insect problems. "Increases in pest problems are less than what farmers in Margaret's survey anticipated," says Carroll. "There was a marked decline in [major] pest problems from the beginning of the reduction (14.2 percent) to the present (6.3 percent). Some farmers [said] they were willing to put up with more weeds and insects now that they have reduced pesticide inputs, but these increases in pests are not affecting yields too much."

Equipment changes are one area where myth and reality matched almost perfectly. More than 45 percent of the conventional farmers fear having to buy new equipment if they cut chemicals, and more than 35 percent say that's a main reason they'd avoid low-input farming. Some 38 percent of the low-input farmers said such purchases are indeed common.

In looking back, both researchers think the word "new" may have confused the issue. Did farmers take it literally—as in "showroom new"? Or did they simply add one or two items—a rotary hoe, for example—that might be bought used? "I think we'd have to take a closer look at what they are buying," says Carroll.

And Some Surprises!

Just 28 percent of the conventional growers expect increased labor needs to be a problem if they cut back on chemicals. But 64 percent of the low-input farmers said it actually is. "A lot of my farmers reported that they do have labor and management increases, but we really don't know where those needs increased," says Carroll.

They don't seem to be caused by more field operations. Only 16 percent of Carroll's respondents said more time cultivating weeds was a major problem when they began cutting chemicals. And only 7.4 percent said it was a major problem after they had completed the transition. The percentages are even lower for more time spent handling manure.

Carroll and Krome think the farmers took planning and thinking time into account when answering labor questions. "One farmer commented that if you asked him where specifically labor inputs increased, he couldn't point to hours logged on a tractor or spent in a barn," observes Carroll. "What he could say is that he was paying more attention to the condition of his crops and fields—where weeds were looking bad, where the corn looked bad, what he should plant in a particular field the following year."

Criticism by neighbors appeared to be another unexpected concern among the low-chemical farmers. Some 21 percent felt peer pressure when they switched to low-input farming methods.

Krome's respondents didn't anticipate such worries. They knew there would be skepticism among neighboring farmers and chemical suppliers, she says. "However, if you ask whether that matters to them, it's not an issue at all." Just 4 percent said peer pressure would keep them from cutting back on chemicals. "Some wrote notes on the questionnaire saying, in effect, 'Who cares?'" observes Krome.

Loss of base acres for government crop programs was a minor worry in both surveys. Only about one-fourth of the low-input farmers said crop base acreage declined when they cut back on chemicals. Even fewer of the conventional farmers (about 3 percent) said this would be a barrier to reducing chemical use. In fact, just 21 percent of the conventional farmers said they'd be motivated to cut back on chemicals if crop payments did not depend on base acreage. "That was the next to the lowest of all the factors that would help them reduce chemicals," reports Krome. Only an additional 1 percent to 2 percent tax on farm chemicals ranked lower.

Bridging the Info Gap

The most confusing responses concerned the issue of reliable information on low-input farming. Just 18 percent of the conventional growers said a lack of reliable advice is stopping them from trying such methods. "And yet throughout the rest of the survey, there are indications that they really don't have the information they need," says Krome.

For example, when asked specifically if they think there are reliable alternatives to insecticides and herbicides, only 30 percent answered yes.

More hints of an information gap emerged when Krome asked where farmers think they could get reliable advice on cutting chemicals. Even the top two choices—successful low-input farmers and Extension specialists—were listed by just 30 percent and 23 percent of the farmers, respectively. The lowest-ranking choice, university experiment stations, was considered reliable by only 10 percent of the respondents. "It's clear that these farmers are not confident in the current sources of information," says Krome.

Carroll agrees. "My [low-input] farmers found information to be much more of a problem than Margaret's perceived it to be," he says. Fully 71 percent of his respondents said they lack reliable advice on maintaining soil fertility and controlling pests without chemicals.

"One of the reasons information may be such a serious problem is related to the sources these farmers use," explains Carroll. "Reduced-chemical farmers are relying on sources outside the conventional information network. In a way, they're setting up their own networks

---

Earth Care Action

Hometown Hero

After his three-year-old son became ill with leukemia, Paul Buxman loaded up all the chemicals from the "poison shed" at his Dinuba, California, orchard and gave them away.

The orchard is in the San Joaquin Valley, an area marked by clusters of childhood cancers. Buxman suspected Wyeth's illness was linked to drinking water contaminated with DBCP, a nematocide banned in 1979.

Within three years, yields from his 40 acres of peaches, plums, grapes, nectarines, and kiwifruit returned to normal.

Buxman traded a $5,000-per-year chemical bill for healthy soil enriched with local, free resources—rice-hull chicken litter and grape and raisin wastes. To control nematodes, he plants California poppies and Cahaba white vetch.

Because he "wanted to help everybody farm without polluting," he founded California Clean Growers in 1987 to help traditional farmers cut chemical use. As chemotherapy saved his son's life, Buxman realizes small amounts of chemicals used with discretion sometimes can be a necessary choice.

Paul Buxman and his son, Wyeth. (Mark Alvis/*New Farm*)

California Clean now has 250 members "with nothing to lose but heavy chemical bills," he says.

From Organic Gardening, *June 1989. Reprinted by permission.*

based on their own experiments, what they've read, and what other low-input farmers have found.

"These alternative information sources are not always satisfactory," says Carroll. In fact, he points out, farmers who rely on them were also most likely to report yield losses and other production problems. "This is not surprising, considering much of their knowledge is gained through trial and error," he says.

Farmers will always complement scientific advice by watching other growers and testing ideas for themselves, observes Carroll. But that's not necessarily bad. "Over time, their information problems decreased," says Carroll. Nearly 71 percent of the low-input farmers said lack of information was a problem when they first cut chemicals, but only 59 percent said it was a problem later.

Still, says Krome, "Farmers have a wealth of knowledge among themselves. They need to communicate and push their public institutions to facilitate the exchange of that information and provide accurate information on specific techniques."

"We'll Never Go Back"

The most valuable information from these surveys had nothing to do with perceived or actual barriers to low-input farming. It concerned reasons why farmers would reduce—or already have reduced—chemicals.

Myth vs. Reality

	Myth (conventional farmers)	Reality (low-input farmers)
Yields	70.9% think yields will fall if they cut chemicals.	35.3% say yields fell; 17.6% say they rose; and 47.1% say they stayed the same.
Profits	63.8% say fear of lost profits is one of top 3 barriers to low-input farming.	76% say lower costs increased profits.
Pest pressure	38% say potential rise in pest problems is one of top 3 reasons they don't cut chemicals.	43.4% say pests were a problem at first; just 33.3% say they remained a problem after a few years of farming with fewer chemicals.
Equipment	35.4% say potential for increased equipment needs would keep them from trying low-input methods.	37.6% say equipment needs increased.
Labor needs	28.3% expect low-input farming to increase their labor needs.	63.6% say labor needs increased.
Information on alternatives	18.2% say lack of good information would keep them from trying low-input methods.	70.8% say finding good advice was a problem at start of transition; 59.1% say it remained a problem afterward.
Crop base	13.1% expect cutbacks in crop base acres if they use fewer chemicals.	25.6% say base acres fell when they cut back on chemicals.
Criticism from neighbors	4.1% say criticism by neighbors would keep them from cutting chemicals.	21.4% feel pressure from neighbors; 59.1% do not.

SOURCE: From the *New Farm*, May/June 1989.

Two factors that would motivate conventional farmers to make such a switch are individual assistance from specialists in reducing chemicals and access to markets that pay premiums for low-chemical commodities. But premium prices didn't seem to be an issue among the low-input farmers. Only 19 percent of them tap such markets. "I didn't ask whether that was a problem, though," says Carroll.

The real motivator for both groups: environmental concerns.

A whopping 84 percent of the conventional growers said they'd likely cut chemical use if studies showed groundwater in their area was contaminated with farm chemicals. Of the low-input farmers, 80 percent said public attention to soil and water problems actually was important in their decision to cut chemicals. "A lot of farmers even wrote about their concern for groundwater pollution," says Carroll.

Economics were the second biggest motivator to the low-input group. About two-thirds cited that as a reason they farm with fewer chemicals.

The obvious conclusion is that farming can be profitable and safe at the same time, says Carroll. "We're used to seeing that as a trade-off, but I think a lot of these farmers would agree that you can do both."

The strongest proof came when Carroll asked his low-input farmers what, if anything, would cause them to start relying more heavily on chemicals. Increased weed and insect problems ranked highest, with 29 percent. Higher commodity prices and lower chemical costs were each mentioned by fewer than 10 percent. So was increased pressure from farm lenders.

The majority—fully 75 percent—said, in effect, "Nothing. I intend to continue farming the way I do now."

"These farmers seem to be committed to what they're doing," observes Carroll. "They like the way they're farming, now, and they're going to stick with it despite those other factors."

From the New Farm *(222 Main Street, Emmaus, PA 18098), May/June 1989. Reprinted by permission.*

Earth Care Action

Hormones to Be Used in Lieu of Conventional Pesticides

By Charles W. Stevens

Synthetic growth hormones are moving closer to market as safer, more effective weapons against insects than conventional pesticides.

S.C. Johnson Wax Co. in Racine, Wisconsin, says it's taking steps toward including in its Raid pesticides a hormone mimic that keeps juvenile fleas from growing into adults, the stage in which the bugs become blood-sucking pests to pets and people. Once trapped in a juvenile state, fleas "just crawl around in the rug feeding on dirt" and soon die, says Roger Grothouse, head of the Raid Center for Insect Control.

The juvenile hormone mimic, fenoxycarb, was recently developed by University of Florida and U.S. Department of Agriculture researchers, along with another substance, diflubenzurone, which interferes with insects' molting. The latter substance kills bugs by preventing them from growing new protective shells after shedding their outer skeletons.

"The chemicals that we've been working on are specific to insects and have little or no effect on mammals," says Phil Koehler, a University of Florida entomology professor. Conventional pesticides, by contrast, attack the insects' central nervous systems and are toxic to pets and humans, as well. Also, current flea-pesticide chemicals break down in sunlight and require monthly applications, while hormone mimics will last up to six months.

Reprinted by permission of the Wall Street Journal, *December 30, 1988. Copyright © 1988, Dow Jones & Company, Inc. All rights reserved worldwide.*

Please Don't Eat the Daisies

By Sharon Begley with Mary Hager

Ah, springtime in suburbia: crocuses pop, daffodils bloom—and lawns sprout signs warning that they have been sprayed with pesticides so humans and other living things should keep off. These unlikely harbingers of the season are now required by law for professionally sprayed lawns in at least six states. Others are considering similar "posting" laws. At the same time, businesses are trying to convince consumers that lawn pesticides are almost as benign as a dandelion salad. "There is such flagrant misrepresentation [that] consumers are unaware the products have risks," says Jay Feldman of the National Coalition against the Misuse of Pesticides. New York Attorney General Robert Abrams concurs. He is suing ChemLawn, the largest lawn-care firm, alleging false and misleading safety claims in its ads.

Some 300 million pounds of pesticides are applied to lawns, gardens, parks, and golf courses each year. Some of these compounds can cause nerve or kidney damage, sterility or cancer in lab animals and workers. But because epidemiological research is so sparse, it's nearly impossible to pin illnesses on lawn products. One closely watched attempt is the case of Navy Lt. George Prior. In 1982, according to his widow, he developed a headache, fever, and nausea after golfing at Virginia's Army Navy Country Club, which was reportedly treated with the fungicide Daconil. He was suffering from toxic epidermal necrolysis (TEN), which makes skin fall off in sheets and causes organ failure. Prior died shortly after, and his widow contends that the TEN came from his exposure to chlorothalonil, an ingredient in Daconil. A lawsuit filed by Prior's widow is scheduled for trial this month [May 1988]. A spokesperson for Daconil's former manufacturer said the claim was "without merit."

Few people are as chemically sensitive as Prior. Yet lately there have been hints that healthy people are harmed by pesticides. "An increasing number who thought they had the flu are finding physicians can document a cause-and-effect relationship [to lawn chemicals]," says New York Assistant Attorney General Martha McCabe. For some the risks are worse. In a controversial study, the National Cancer Institute found that Kansas farmers who apply 2,4-D, a weedkiller used in 1,500 over-the-counter products, had a risk of non-Hodgkin's lymphoma, a rare cancer, up to seven times higher than average.

Now there is also concern over the solvents, dyes, and other "inert" ingredients in pesticides. Of the 1,200 inerts, the Environmental Protection Agency (EPA) considers 300 safe—and about 60 of "significant toxicological concern." Among them: benzene and chloroform. Both cause cancer in animals. Unlike active ingredients, inerts do not have to be listed on pesticide labels. But EPA has announced that the 60 most hazardous inerts must be removed from pesticides by October or be named on labels.

Clearly, killer lawns aren't sweeping suburbia. Roger Yeary, chief of health and safety at ChemLawn, argues that few homeowners get the "repeated exposure to a constant, low-level amount" of a chemical that produces chronic effects, since most lawn compounds quickly decompose. ChemLawn uses about 25 pesticides. In a brochure for customers, it stresses that none are "known or probable human carcinogens" and are "practically nontoxic." Yet one of its fungicides, mancozeb, breaks down into a compound EPA categorizes as a known human carcinogen. Another, chlorothalonil, causes cancer in animals. ChemLawn says it uses the fungicides hardly at all. Cancer, of course, is hardly the only concern. Data on the neurobehavioral effects of long-term, low-level exposure to lawn pesticides are poor or nonexistent.

Safe as Baby Aspirin?

Many consumers aren't aware of these doubts about safety. One major reason, according to a recent report by the U.S. General Accounting Office (GAO), is "false and misleading" safety claims by pesticide manufacturers. GAO's undercover team also found that professional applicators made unsupportable assertions, such as saying products "absolutely cannot harm children or pets." The New York suit cites ChemLawn's claim that a child would have to eat almost ten cups of treated lawn clippings to equal the toxicity of one baby aspirin. In fact, the danger is not that people graze on the lawn but that they inhale fumes or absorb residues through skin.

Although most of the furor—and the legal action—is aimed at the $2 billion lawn-care industry, do-it-yourselfers may be more dangerous to themselves and others. They can be exposed to concentrated pesticides—which they mix with water, for instance—and not

---

Earth Care Action

Lawn-Care Companies Now Offer Alternatives

Concern over the potential hazards of chemical lawn treatments has spawned a growing number of lawn-care companies offering pesticide-free options. They may not produce a lush, picture-perfect yard, but they usually provide an improved lawn and greater peace of mind for worried homeowners who prefer to go organic.

Participating branches of Barefoot Grass Lawn Service, ChemLawn, Lawn Doctor, Spring Green Lawn Care, and Tru Green now offer low-pesticide, pesticide-free, or fertilizer-only treatments—some only on request. Smaller firms such as Hydro Lawn, with branches in four eastern states, will also design pesticide-free treatments.

Local lawn-maintenance firms may provide these services, too. For example, Living Resources Co. of Citrus Heights, California, offers a completely organic landscape-maintenance program, including an initial consultation, periodic fertilization, and organic pest control. A season of treatments for a quarter-acre lot costs around $225, probably more than the cost of a season of traditional care (typically about $200 from one well-known firm).

If you have no success finding a cooperative service in your area or would like to treat your lawn yourself, Biological Urban Gardening Services (BUGS), a fledgling network of organic landscapers directed by the founder of Living Resources, may be able to help. For more information, write to BUGS, P.O. Box 76, Citrus Heights, CA 95611. Another source of information is the National Coalition against the Misuse of Pesticides, 530 7th Street SE, Washington, DC 20003.

From Changing Times, *August 1987. Reprinted by permission.*

merely to the dilute versions that go onto lawns. Also, people treating their own lawns are notorious for overdosing. By one estimate, pesticides applied at a rate of two pounds per acre in farmers' fields go on at ten pounds per acre on lawns. A compound that is relatively safe at low levels may not be so benign in the amounts favored by homeowners. Yet noncommercial sprayers aren't covered by posting laws. "What's the sense of exempting homeowners...but restricting licensed, trained lawn-care professionals who adhere to our association's code of ethics?" asks James Wilkinson of the Professional Lawn Care Association of America. Perhaps the best advice to pesticide users is, know what you're using, don't use it if you don't have to, and don't regard a perfect lawn as the sine qua non of suburban bliss.

From Newsweek, *May 16, 1988. Copyright © 1988, Newsweek Inc. All rights reserved. Reprinted by permission.*

PLASTICS

Persistent Peril

By Pat Wittig

Pollution Solutions 78
Plastic and You 80
American Myopia: Focusing on Short-Term Profits 83
Most Throwaway Plastics Face a Ban in Minneapolis 85
A Second Life for Plastic Cups? Science Turns Them into Lumber 87
Diaper Wars! 89
Comparing Bottom Lines 90
Diaper Solutions 93

It began, innocently enough, in 1868 with the development of celluloid billiard balls and eyeglass frames. The Age of Plastic had arrived. A century later, the growth rate of plastic production was called "phenomenal" by the Environmental Protection Agency, as U.S. industry yielded 16 billion pounds of the light yet lasting material.

If that was phenomenal, then the 53 billion-plus pounds of plastic resins sold in the United States in 1978 must be considered awesome—or, perhaps, awful to those of us concerned with the health of our planet. A glut of plastic waste now threatens marine and beach environments, world wildlife, and landfill space, and plays a major role in our escalating municipal taxes.

Can plastic be blamed for so many woes when it constitutes only 7 percent of the solid waste stream? Yes it can, when that 7 percent is viewed as more than 1,000 pounds of lightweight, essentially nondestructible garbage dropping from the hands of each U.S. citizen every year.

Plastic's low density makes it more voluminous than other types of garbage. Several consulting firms report that compacted plastic packing waste has a density of about seven

From Organic Gardening, *February 1989. Reprinted by permission.*

pounds per cubic foot. So plastic's 7 percent by weight translates to 25 percent to 30 percent by volume of the municipal solid waste stream.

While industry spews out more and more plastic, the burden of safe disposal and recycling of these products mainly falls on individual communities, entrepreneurs, and concerned legislators. The move toward new recycling technologies and new, degradable plastics may never outpace the flow of plastic waste unless the industry itself confronts the problem more directly.

"The component that is missing," says Rob Stuart, lobbyist for the New Jersey Public Interest Research Group (PIRG), "is the idea that the manufacturing sector or the distribution sector plays any role in getting materials back or even that they should make their products more recyclable." Stuart and others who favor recycling call for industry regulations, taxes on containers, deposit/refund laws—and even bans—as additional measures to reduce municipal wastes.

Indeed, the trouble with plastic is its persistence. A plastic motor oil container tossed onto a roadside, dumped in an ocean, or trucked to a landfill will remain essentially unchanged for several hundred years. It is difficult to compress, toxic to incinerate, and does not decay. It just adds to the heap. Plastic pollution is truly a problem that does not go away. But it does ebb and flow like the tides—literally—for the low density of plastic allows it to float.

In fact, the most publicized concerns about plastic trash are its effects on marine environments. Heart-wrenching photos are increasingly common—of dead sea turtles with plastic bags hanging from their gaping mouths, of sea birds hobbled and harnessed by plastic six-pack yokes or else dangling from tangled fishing line, of seals with plastic rings clamped around their muzzles or biting into their necks.

Although no one knows exactly how much garbage is ocean-bound, environmentalists estimate that the amount of plastic dumped overboard reaches more than one million tons per year worldwide. More than five million plastic containers reportedly are dumped every day from the 50,000 merchant, naval, fishing, and pleasure ships that ply the seas.

In addition, commercial fishing fleets from Japan, Taiwan, and Korea daily lose about 12 miles of the more than 20,000 miles of plastic drift nets that are cast every day during the five-month fishing season. These lost and discarded plastic nets and lines are efficient killers, attracting birds and mammals—"incidental takes," as they are called—to feast on the bounty of trapped sea life, only to become trapped themselves.

A great number of sea birds, marine mammals, fish, and turtles—as many as 100,000 marine mammals per year, according to environmentalists' estimates—lose their lives, in agony, when they ingest plastic debris, apparently mistaking it for food. Pellets meant for use in manufacturing plastic products, as well as plastic foam cups, plastic caps, rings, straps, toys, and disposable razors and lighters, are just a few of the items retrieved from the gullets of dead sea birds, fish, and mammals.

Ocean dumping, particularly of plastics, is a global concern. The United Nations 1973 International Convention for the Prevention of Pollution from Ships included a provision, called Annex V (five), that prohibits ocean disposal of plastics, limits ocean dumping, and forces ports to provide trash facilities for ships. The United States finally ratified Annex V in late 1987. At the same time, it passed a federal law (The Plastic Pollution Research and Control Act of 1987, which took effect December 31, 1988) banning all dumping of plastic materials in the ocean.

Ratification by the United States means that enough nations, representing more than half of the world's commercial shipping activities, support Annex V, but it is not a panacea. It does not address incidental takes or lost drift nets. Nor is it fail-safe, because laws of the open sea are difficult to monitor and enforce.

Military boats and ships are not now required to comply with Annex V except as decreed by each nation. U.S. military vessels, which daily dump overboard four tons of plastics, including 450,000 plastic containers, have five years to begin complying with Annex V and the 1987 U.S. law. The U.S. Navy is researching solutions to its plastic waste problem, reports Kathy O'Hara, a plastics expert with the Center for Marine Conservation.

"They've been canceling supplies such as plastic cups and garbage bags. And they're trying to minimize the other types of plastic they use," O'Hara adds. "But we definitely can't rest now and say, 'Good! We have a law. Now everything's all taken care of.'"

Plastics-related deaths of wildlife in forests and fields are not as well documented. Occasionally, accounts of deer choking on plastic bags and similar gruesome stories appear in local newspapers; yet, on land, the problem of

---

Earth Care Action

Pollution Solutions
By Matt Damsker

As the saying goes, "Build a better mousetrap"—or, in this case, craft a free composter and strike a blow against the plastic waste stream.

That's what's happening in Canada's Peel Region, where local Boy Scouts have found a creative way to help local industry rid itself of empty plastic drums without landfilling them. The scouts simply cut both ends off the barrels and drill 16 two-inch holes in each one, transforming them into home composters.

The scouts receive $10 from the region for each unit, and the composters are distributed free to interested homeowners. Glen Milbury, Peel's industrial waste reduction coordinator, reports that 3,000 of the composters have been distributed so far. "We've even had to go elsewhere, down into the States, for more drums!"

Meanwhile, farmers and home gardeners are turning more and more to plastic mulch film for earlier germination and increased yield, often using as much as 200 to 500 pounds of film per acre, adding considerably to the plastic disposal problem. Use of degradable plastic mulch is increasing, and recently a Miami, Florida, firm called Ideamasters Inc. introduced a plastic mulch film called Plastigone, which is photodegradable and comes in five different formulations that decay at a controlled rate keyed to specific regions.

Upon exposure to sunlight, Plastigone undergoes stages of degradation—from "embrittlement," in which the film maintains its protective form but is easily broken up by tilling or cultivating, to total disintegration, leaving behind no chemical residue. Ideamasters spokesman Dick Weatherford says that Plastigone is the only time-controlled degradable plastic film on the market. It's available by direct mail to growers. For more information, call (305) 274-1233.

From Organic Gardening, *February 1989.*

Products made from recycled plastics are a boon to recycling efforts. This windbreaker was fashioned from recycled plastic soft-drink bottles by Chima Inc. of Reading, Pennsylvania. (Rob Cardillo)

plastic waste is most often presented as the acme of a mounting waste-disposal crisis that is beginning to scrape the sky and fray municipal purses.

According to Homer A. Neal and J. R. Schubel, authors of the book *Solid Waste Management and the Environment,* waste of *all* types is generated in the United States at the rate of 1.1 billion pounds per day. Landfills are closing as they rapidly reach capacity or when residual environmental problems arise at dump sites and municipal trash incinerators. Fears of surface and groundwater contamination, gas pollution, and traffic hazards caused by an influx of garbage trucks forestall approval of new disposal sites. Growing populations and increasing per capita waste generation compound the problem of shrinking landfill space. It's no wonder: The United States is the world's largest producer of trash.

Without considering industrial waste, consumer waste alone—the stuff that disappears from our curbs each week—is expected to increase to about 3.7 pounds per person each day, or 1,350 pounds per person annually, during the next decade. And the amount of plastics used in packaging and manufacturing is expected to continue to increase. Already, plastic beverage containers, unheard of until the 1970s, now dominate the container industry and litter our roadsides and beaches.

The most touted method of dealing with plastic trash is recycling. There are two broad categories of plastics: thermoplastic, the kind used in most packaging, which is reusable because it can be reheated and reformed several times without degrading, and thermoset plastics, which cannot be reformed due to permanent chemical changes in the material once it is heated.

The plastics industry itself has been recycling much of its "in-house" thermoplastic scraps and waste. Mobil Chemical, for instance, purchases 20 to 33 percent of all industrial polyethylene scrap and uses it in the production of its "Hefty" trash bags. That activity alone accounts for about 100 million pounds of plastic that is recycled each year before it leaves the factory.

Attention to recycling in the press and in state legislation, though, is focused mostly on postconsumer wastes, 150 million tons of which were carted from our curbsides in 1985, with plastics accounting for 10.5 million tons of it. Some plastics industry representatives call that an insignificant amount, but at seven

---

Earth Care Action

Plastic and You
By Pat Wittig

Here are some easy ways you can reduce your household and community flow of plastic waste.

• Request paper bags at the supermarket, or use your own fabric or string net sacks at the market. You also can ask your local market to order and use degradable plastic bags from Beresfore Packaging Inc., 155 Miles Standish Boulevard, Tauton, MA 02780; (508) 824-8636.

• Reuse plastic bags several times—for wrapping sandwiches, storing leftovers, or freezing vegetables.

• Seek out a recycling operation that will accept your plastics. Suggest plastic drives when schools propose paper drives.

• Support bottle-bill legislation that taxes or requires a deposit on plastic containers.

• Buy foods in bulk form or join a food co-op to avoid excess packaging.

• Buy products in glass, aluminum, or cardboard containers and use wood, leather, wool, cotton, and other natural products whenever possible.

• If you use plastic trash bags, plastic mulch, or disposable diapers, ask your market to stock degradable varieties.

• Request paper containers at stores that serve food and beverages in plastic foam.

• Suggest recycled plastic products such as curbs, speed bumps, decks, etc., when your town is planning to build parking lots, parks, or recreation areas.

• Support legislation restricting multi-layer packaging, especially in single-use products.

From Organic Gardening, *February 1989.*

pounds per cubic foot, that's enough compacted plastic to fill 125,000 four-bedroom houses, floor to ceiling, each year.

What may be insignificant, some environmentalists point out, is the fraction of that total that is receiving most of the attention of recycling efforts: plastic beverage containers, especially bottles made of PET (short for polyethylene teraphthalate, a type of polyester).

Significant or not, PET bottles accounted for 24 percent of the 2.9 billion pounds of plastic bottles made in the United States in 1986. These containers now are assigned entire aisles in supermarkets, capturing the fancy and finances of recycling entrepreneurs. Denser PE (polyethylene) plastics, familiar as milk and detergent bottles, films, and base cups of PET bottles, are also of interest to recyclers but result in a less valuable end product. In 1986, PE accounted for about 62 percent of the plastic bottles produced in this country.

The plastics industry can answer critics who charge that it isn't pulling its weight on the solid waste issue. According to Susan Vadney of the Society of Plastics Industry (SPI), the industry is "demonstrating its commitment to sound environmental policies" through such programs as its voluntary coding system for plastic containers, which identifies them by resin type for easier recycling.

The industry also funds recycling technology through the Center for Plastics Recycling Research (CPRR) at Rutgers University in New Jersey and other recycling and reprocessing projects. At the same time, the industry supports research on incineration of plastics—not a viable solution as far as environmentalists are concerned—and typically resists "bottle-bill" legislation, which mandates the return flow of used beverage containers for recycling.

Plastics recycling technology still is in its infancy. Research facilities such as CPRR are gaining momentum, however, and are willing to pass on their knowledge to the many communities, industries, agents, and others that now are seeking to establish plastics recycling operations.

Briefly, the recycling process involves chopping the plastic bottles—caps, labels, and all—into fingernail-sized chips, washing them, and separating the PET from the PE (PET floats, while PE sinks in water). The clean, separated plastics are then packed for shipping to manufacturers to begin their second coming as thermal jacket fillings, flower pots, new base cups, or other items that by law will not be in contact with food.

Another recycling strategy, based on the extrusion-molding methods used in Belgium, is an increasingly common alternative to landfilling or incinerations. Extrusion molding uses PE as well as commingled, or mixed and contaminated, plastic wastes that were previously considered impossible to recycle because the components are as incompatible as oil and water.

The end product of extrusion-molding recycling is "plastic lumber." Although plastic lumber would never fool a woodpecker, it is similar to wood in that it can be sawed, planed, drilled, or nailed with standard carpentry tools.

"It is not quite as strong as wood because it has no fibers," explains Jose Fernandez, plant manager of Rutgers' CPRR pilot operation. It is practically indestructible, though, because it is impervious to water, insects, bacteria, fungi, chemicals, and corrosion. The product may be used to build fences, platforms, piers, park benches, playground equipment, or any items (except buildings) that conventionally are made from wood.

Recycling to produce plastic lumber eliminates the problem and expense of separating plastics by type, and the finished products are highly marketable. Witness the success of companies such as two-year-old Polymer Products in Iowa Falls, Iowa. Inspired by the success of similar companies in Europe, Polymer Products' founders make and sell traffic control products and livestock feeders and platforms from recycled plastics.

Greg Mattson, vice president of the company, says the possibilities of new plastic products from old are "almost endless. Most of the material we get is postindustrial waste. But we're looking to increase our postconsumer business." So far his firm has had few dealings with the public sector.

In New Jersey—currently the only state in which statewide recycling is mandated by law—about 80 of the more than 400 communities now collect plastics for recycling. That number is increasing, says Aletha Spang, acting administrator of the N.J. Department of Environmental Protection's Office of Recycling, although plastic is not one of the materials that by law must be recycled.

New Jersey's communities can choose three of four categories of recyclables—metals, glass, paper, and yard wastes—to cut municipal solid wastes by 25 percent. "So they get the aluminum, glass, and paper, and they're off the hook," says Fernandez. "But the volume of plastics in the waste is tripping them up."

Since dumping fees at landfills and incinerators are frequently assessed by volume, it makes sense that more communities will try to control rising municipal taxes by jumping on the plastic recycling bandwagon. As oil prices rise, too, interest in plastics recycling will likely increase. "Plastics represent some percentage of every barrel of oil. They are valuable resources that should be mined from the garbage stream," points out Christine Latella, director of recycling for Mobil Chemical.

Obviously, media attention to the plastic waste problem can only raise the consciousness of consumers. Some, like Karen Schneider, a working mother of three who lives in New Jersey, now refuse to buy detergent, milk, or juice drinks in plastic containers until their communities include plastics in their recycling plans. "If it can't be recycled, then it shouldn't be made," Schneider asserts.

Recycling and other measures are not perfect solutions to the plastic waste problem. At their current rate, they may not even dent the amount of plastic waste destined for landfills. Another solution recently proposed is to change the indestructible nature of plastics so that they will decay in landfills, in the ocean, or on the roadsides.

To this end, researchers are developing plastics that degrade when exposed to sunlight, moisture, or microbes. The basic method for accomplishing this was actually stumbled upon in the mid-1970s when British university researchers, searching for a way to give plastic bags a silkier texture, added cornstarch to a plastic base material.

The realization that such bags decay did not hit home, however, until ten years later. That's when researchers at the St. Lawrence Starch Company in Toronto and others developed the concept and began to market biodegradable trash bags, plastic mulch, and water-soluble laundry bags for hospitals. Since then, degradable diaper liners, six-pack yokes (required in 16 states), and other products have been introduced. The time it takes for such items to deteriorate—a few months to 15 years—is controlled by adjusting the proportion of cornstarch, lactic acid, or other natural polymer, to plastic: The more starch, the faster it decays.

Representatives of the plastics industry have raised objections to starch-based biodegradable plastics, saying they are too fragile. Purdue University researcher Ramani Narayan has developed a technique that he says overcomes the problem. By using "technology that the polymer industry is more comfortable with," says Narayan, natural polymers such as starch are combined with synthetic polymers (plastic), making materials that are as strong as conventional plastics yet break down in the presence of "garbage microbes."

Narayan's technique of copolymer bonding for this purpose is still in the developmental stage. But when it is ready for commercial application in a year or so, "it should be readily accepted and competitive, price-wise," he says. Without this type of "second-generation technology," he continues, "we will be buried in plastics."

Are efforts to control plastic waste too little, too late? "For plastics, it is too little. I don't think it's too late," responds Jeanne Wirka, solid waste expert for the Environmental Action Foundation, a Washington, D.C.-based organization. "The plastics industry has a lot of catching up to do."

To make real progress in reducing the waste stream, says Wirka, "the packaging industry needs to start designing single-material packaging with waste disposal and recycling in mind." She believes this goal is attainable with enough legislative pressure and environmental education.

In West Germany, for instance, the government has the authority to ban any package that is not easily recyclable, refillable, or degradable. Coca-Cola has responded to that by making a bottle there that is all PET, as opposed to the PET/PE bottle sold in the United States. Denmark, Norway, Japan, and individual communities worldwide have similar restrictions. Florence, Italy, has banned all plastic food containers from sale in that city.

You may notice that many products packaged in Europe and sold here are packed in glass, metal, or paper, and that the cardboard and plastic parts of the packages are instantly separable, not bonded permanently as they usually are in American-style packaging. It's time that we heed this implicit message and pay attention to the problem of plastic pollution. If not, we may someday be viewing our natural world through a plastic tunnel with less and less light at the end.

American Myopia: Focusing on Short-Term Profits

By Ellen Kunes

In a darkened auditorium, a division director from the Food and Drug Administration (FDA) clicks her way through a series of slides. Each one depicts the food containers of the future—an endless array of reheatable plastic pouches, trays, bags, bowls, and cylinders—stuffed with assorted baby foods, soups, whole dinners, and desserts. "What used to appear in metal cans is now likely to come in plastic pouches," the FDA's speaker tells us. "These packages may be a panacea for the troubles of time crunching. By 1991," she beams, "we project that 4.7 billion of these containers will be sold."

Then the speaker—just one of several government scientists touting new products at the annual FDA/U.S. Department of Agriculture Food Safety and Nutrition Conference—finishes her presentation, and a reporter rises to question her. "Are any of these new plastics that you've shown us biodegradable?" he asks.

"Not that I know of," she replies.

"Well," he continues, "isn't your agency worried that introducing all of these nonbiodegradable containers in the next few years will worsen the nation's already monumental waste-disposal problems?"

"That's not our concern," the FDA scientist tells him. "This agency's priority is to make foods safe and convenient for today's consumer. The environment is not our bailiwick."

I sit back in my seat then, wondering just whose bailiwick it might be. Later, a few calls to the Department of the Interior and the Environmental Protection Agency reveal that no official government groups are actually monitoring the flow of new plastics into the environment. Given the desperate state of the environment, it seems clear that we no longer have the luxury of this sort of tunnel vision. We simply can't afford to ignore the long-term effects that our new inventions might have on our nation—and on the wider world as well. Indeed, as Mary Catherine Bateson, anthropologist and author of *Thinking Aids* sees it, we Americans have always tended to ignore future concerns in order to reap immediate benefits. "We have not learned to make the necessary connections between what we do today and the results of those actions on our society in a generation or two," she says.

Instead, our myopia compels us to live in a constant state of crisis management. For years environmentalists warned that our system of landfills and incinerators could not continue to handle the rapidly increasing waste stream. They made dire predictions that by the year 2000 most of the nation's landfills would be packed to capacity. But not until the effluvia began to wash up on our shores, not until we saw our garbage scows sailing the oceans in search of dumping grounds or were confronted with the necessity of building incinerators in our own backyards, did we acknowledge that we had a problem.

Seeking short-term profits has caused many other environmental messes, from acid rain and water pollution to the greenhouse effect. "To truly solve these dilemmas," says Bateson, "we must make our children understand the constant care ecosystems require in order to survive."

Bateson likes to describe the benefits of taking the long view by comparing the handling of the environment to a strategy employed by a champion dogsled racer. For the past three years Susan Butcher has won the Iditarod, a grueling two-week, 1,000-plus-mile race across Alaska. "The way she wins is by caring for her dogs and sustaining their strength," says Bateson. "The men who compete against her always seem to be charging on ahead, but toward the end of the race, their dogs become too tired, while Butcher's team—which has been more carefully paced throughout the race—overtakes them. Butcher is an expert on what environmentalists call sustainability. If the time frame you're looking at is long enough, you are not rewarded for wearing out your dogs or for exhausting the soil, burning the rainforests, polluting the rivers. You win only when you care for these things consistently."

Is it really possible to expect people to eschew convenience—in this instance, to reject the new wave of polluting plastics—in favor of a cleaner world 100 years hence? Bateson believes that our success in this depends on our ability to connect the future to ourselves—that is, to envision how the choices we make today will affect the sort of world our children and grandchildren will have to live in. "We must understand," she explains, "that each of us will be personally injured—through our progeny—if we don't expand and extend our vision of the world."

From Omni, *March 1989. Reprinted by permission.*

Earth Care Action

Most Throwaway Plastics Face a Ban in Minneapolis

By William E. Schmidt

In an effort to reduce a growing tide of plastic waste it must either burn or bury, the Minneapolis City Council has adopted an ordinance to ban most throwaway plastic food packaging from grocery store shelves and fast-food outlets.

The legislation, which does not take effect until July 1990, would permit the use of items like plastic milk jugs or soft-drink bottles only if the city can work out an acceptable recycling program for plastics.

The Minneapolis ordinance and a nearly identical piece of legislation pending before the St. Paul City Council are the most far-reaching efforts yet by local governments to restrict the growing volume of plastic waste making its way into the nation's municipal landfills and incinerators.

In general, the ordinance would prohibit any kind of packaging that is not environmentally acceptable: that is, packaging that is not naturally degradable, like paper; returnable for reuse, like some glass containers; or acceptable for recycling, like metal cans.

Long Island Ordinance

Until now, the most ambitious attempt by local officials to restrict the use of plastic packaging was an ordinance adopted in Suffolk County, Long Island, that barred the use of polystyrene plastic for fast foods or other foods packaged by local retailers. The ordinance is scheduled to take effect in July [1989] unless legal challenges succeed.

The ordinances here cover the same items as the Long Island legislation, as well as plastic containers used for milk, peanut butter, ketchup, syrup, and other items. It would exempt some plastics, like the clear wrapping used to cover meats or microwave dinners, for which there are no easy alternatives.

Advocates of the ordinance say that the plastics take up too much space when buried and do not decompose. They add that, when burned, the plastics might release harmful emissions.

The Minneapolis ordinance was strongly opposed by groups, both local and national, representing food retailers and manufacturers, soft-drink bottlers, and plastics companies. Their representatives argue that the law will raise grocery prices. They also say that the environmental problems attributed to plastic waste have been overstated.

The city currently collects glass, paper, and aluminum as part of a curbside recycling program. Steve Cramer, the sponsor of the ordinance, said the cost of recycling plastic was prohibitive, without help from industry, because only about 20 percent of the plastics used in packaging can be recycled, given current technology.

Grim Estimate on Trash

The Environmental Protection Agency has estimated that such trash will make up 15 percent of the nation's solid waste by the turn of the century, more than double what it is now.

The ordinance was passed unanimously by the city council. Some members remarked that they had been deluged by hundreds of telephone calls from consumers, urging them to approve the ban.

The flood of support came in response to a series of radio and newspaper advertisements placed by opponents of the ordinance, including grocers and plastic manufacturers, asking consumers to do just the opposite—to call their council representatives and speak out against the ordinance.

"The people have clearly spoken," said Dennis Schulstad, the city council's lone Republican member, who, until this week, was undecided about the ordinance. "I had more calls on this one issue than anything that has ever been before the council."

Since the ordinance does not take effect for another 15 months, opponents hope that the Minnesota state legislature will pass a bill prohibiting cities from imposing such restrictions on food packaging.

Opponents argue that the state can and should preempt such local laws until the legislature can institute a uniform statewide policy on recycling and solid waste disposal.

If the legislature fails to act, litigation would be the likely next step, according to Joel Hoiland, who is president of the Minnesota Grocers Association, which represents about 1,300 retail and wholesale grocers across the state.

Hoiland said the Minneapolis and St. Paul ordinances would drive up grocery prices and push consumers to suburban stores that would not be subject to the same restrictions.

He said one local food store chain has estimated it would cost an additional $1.5 million per store to substitute glass or paper packaging for plastic.

Roger Bernstein, director of state government affairs for the Society of Plastics Industry in Washington, described the proposal here as "insane," and said it would have a devastating impact on grocers, retailers, and eventually, consumers by driving up costs.

"I would have to harken back to Prohibition to think of a precedent," Bernstein said. "We don't even ban unsafe products in this country."

From the New York Times, *April 1, 1989. Copyright © 1989 by the New York Times Company. Reprinted by permission.*

Earth Care Action

A Second Life for Plastic Cups? Science Turns Them into Lumber

By Robert Hanley

A new park bench made of sturdy plastic sits in the hallway of a research office at Rutgers University in Piscataway, New Jersey, an ornamental testament to a useful second life for the plastic-foam cup and hamburger box.

These cups and boxes, the throwaway staples of the fast-food industry, together with supermarket and school cafeteria trays, make up 40 percent of the plastic "lumber" used in the bench. The rest is from plastic bottles, melted and remolded, that once held such things as anti-freeze and detergent, yogurt, liquor, and margarine.

For the last year, scientists at the Rutgers Center for Plastics Recycling Research have worked on blending these plastics into planks and posts that someday, they hope, will become an economical substitute for wood in decks, picnic tables, piers, and fences.

A Race against Bans

The experimental research is something of a race for the center, which is a creation of the plastics industry. Its staff wants to prove that polystyrene—the dominant plastic used in fast-food containers—has value when recycled. The motive is to curb any widening of sentiment among environmentalists and politicians to ban polystyrene from use as a food container.

Technology already exists for recycling the more valuable types of plastics used in soda bottles and in clear jugs for milk, water, and fruit juices.

Before tackling polystyrene, the center helped perfect the machinery that shreds those plastics into flakes, washes them, and separates them from remnants of bottle caps and paper labels.

But recycling of plastics lags far behind newspapers and glass bottles. Widespread use of plastic packaging is only about a decade old, and the public, the industry says, has little awareness that discarded plastic containers are recyclable. As a result, economical methods of collecting and sorting discarded plastics are all but nonexistent.

"We can't get anywhere near enough soda bottles," said Dennis Sabourin, vice president of Wellman Inc., the New Jersey-based leader of the plastic soda-bottle recycling industry.

Deposit Laws in Nine States

Of the 750 million pounds of plastic soda bottles used and thrown away in the country annually, only 150 million pounds, or about 20 percent are recycled, Sabourin said. Nearly all come from the nine states that have deposit laws for plastic bottles: New York, Connecticut, Massachusetts, Vermont, Maine, Delaware, Michigan, Oregon, and Iowa.

Wellman handles about two-thirds of that volume, he said, transforming the flaked plastic into polyester fiber used in wall-to-wall carpeting and furniture cushions and as insulating fill in comforters and ski jackets.

Besides fibers, chips of soda-bottle plastic, polyethylene teraphthalate, more commonly called PET, can be reused to make bathtubs and shower stalls, boat hulls, electrical wall sockets, and plastic panels for cars.

Uses for chips of milk-jug plastic, known as high-density polyethylene, or HDPE, include toys, trash cans, flower pots, pipes, and pails.

Still, the industry estimates that 99 percent of all discarded plastic containers wind up in landfills, in large measure because only soda bottles and milk jugs have proven new uses. Because the containers are, in essence, bulky plastic bubbles, they use considerable landfill space. Hence the value of the research into plastic lumber at the center here.

Ten benches matching the center's hallway ornament were recently shipped from the center to Palm Beach, Florida, the end product of junked plastic packages gathered there last summer in a beachcombing bee.

Piled in the center's workshop are eight-foot plastic planks and posts produced by a Belgian-designed machine that melts plastics to a pudding texture and squeezes it into metal molds.

Will Nails Pop Out?

"These are all prototypes. We're still experimenting with a variety of mixes of plastics that scientists once thought were incompatible," said Dr. Thomas J. Nosker, manager of the project.

Questions still unanswered about the "lumber" include its durability and ability to hold fasteners over years.

"Will there be warping or discoloring?" said Ted Kasternakis, the project's research engineer. "How will the sun change the color? Will there be chipping or cracking? It's all new territory and we have to get a handle on it."

The center was formed in 1985 by the Plastics Recycling Foundation, a nonprofit industry organization. It provides 42 percent of the center's $2.3 million budget. New Jersey provides 30 percent, Rutgers 19 percent, and the rest is from the National Science Foundation and other states.

But even with the center's innovations, officials say, plastics faces another recycling hurdle: lightness—the quality that made plastics an economical replacement for glass containers.

Unless systems are created to sort containers by plastic type, compact them, and collect them, the cost of carrying the bulky containers could jeopardize the value scientists may create for recycled products.

For example, discarded plastic soda bottles are worth three times as much as shattered glass bottles. But a standard garbage truck can carry about 30,000 pounds of crushed glass, worth $600, and only about 1,000 pounds of unshredded plastic bottles, worth $60, according to Guy Watson, manager of technical assistance in New Jersey's Office of Recycling.

"That's the crux of the problem to municipal officials," Watson said. Only 80 of New Jersey's 567 towns recycle plastics, he said.

From the New York Times, *February 22, 1989. Copyright © 1989 by the New York Times Company. Reprinted by permission.*

Diaper Wars!

By Jeanne Wirka

Nobody wants to deal with dirty diapers. "Diapers are filthy, smelly, and no fun," remarks one mother of two with another on the way. "I wish they would go away."

The desire for a disappearing diaper has precipitated a revolution in American diapering practices. Over a few decades, cloth diapers, rubber pants, and duck-headed diaper pins have been all but replaced by Pampers, Huggies, and other single-use, throwaway diapers.

Of course, throwing diapers away didn't make them disappear. The 18 billion disposable diapers Americans throw away each year—enough to fill an Islip-style garbage barge every six hours—are choking up the nation's rapidly filling landfills and increasing the need for other disposal options like incinerators.

Americans are notorious garbage generators. Each of us produces roughly three pounds of trash a day, a full pound more than our West German and Japanese counterparts. And our habits are getting worse. The 160 million tons of municipal solid waste that Americans generated in 1988 will rise to 192 million tons by the year 2000. Meanwhile, 27 states are expected to run out of landfill space in only five years.

As the nation's garbage problem spins out of control, a call for a return to reusable cloth diapers is being heard from growing numbers of environmentalists, waste management officials, and public health professionals.

"It's time to potty train society," says Carl Lehrburger, a recycling specialist with Energy Answers Corp., a solid waste consulting firm in Albany. "The bottom line is that single-use diapers should not be going to landfills."

In the emerging diaper war of cloth vs. disposable, makers of disposable diapers will try to capture the flag on the grounds that their product is more "convenient." But looked at from a diaper-changer's point of view, this factor really ends up in more of a tie. Yes, cloth diapers have to be washed, but not if you get diaper service. And—though most moms and dads don't do it—you're supposed to rinse out baby poop from the throwaway before sending it on its trashy way.

Baby comfort and diaper rash? Some parents swear by cloth, others swear by their disposables. Actual diaper changing? There's really not much difference, time-efficiency-wise. Also on the efficiency front, disposables have to be lugged home from the market week after week. On the other hand, if you bring your cloth-diapered baby to visit grandma, baby's dirty diaper comes home with you.

While the convenience debate ends in a draw, there is a clear difference in two areas—costs to consumers and costs to society. In both regards, cloth wins hands down.

Diaper services—companies that pick up dirty diapers from your home and deliver clean ones—claim that the new national awareness of waste is having an impact on parents' diapering practices.

"We've noticed a pretty significant turnaround in our business in the last year," reports Jack Schiffert, president of the National Association of Diaper Services (NADS). "We feel quite strongly that the growing environmental awareness is responsible."

Schiffert is not alone in predicting a turnaround. The February 1989 issue of *Life* magazine predicts the demise of disposable diapers and advises its readers, "Invest your money in diaper services, because the environment is crying for a change."

Since they began to catch on in the 1960s, disposable diapers have virtually taken over the market. Today, they account for over 75 percent of all diaper changes in the United States and bring in $3.3 billion a year.

Like many of the statistics hurled in the emerging diaper war, figures on how many diapers there are in the municipal solid waste stream are varied and hotly disputed. Procter & Gamble—maker of Pampers and Luvs, and controller of 50 percent of the U.S. disposables market—has arrived at a figure of 0.5 percent, based on a dry weight of roughly 1.33 ounces.

Kimberly-Clark—maker of Huggies, with a 30 percent market share—puts it at 0.6 percent.

But critics point out that used diapers aren't generally dry: virtually all contain urine, and roughly 30 percent contain feces.

A new report released in February by Carl Lehrburger, *Diapers in the Waste Stream,* included waste calculations based on an average "wet" weight of seven ounces per diaper. Lehr-

Comparing Bottom Lines

	Disposable Diapers[a]	Diaper Service	Cloth Diapers and Home Wash
Number of times diaper is used	1	150[b]	90
Cost over diapering period	$1,533 (Average cost of 21¢ per disposable[c] × 7,300 diapers)	$975 (Average cost of $7.50 per week[d] × 130 weeks)	$283 (7 dozen diapers at average cost of $9.23 per dozen[e] × 130 weeks = $64.61 Home laundering[f] cost of 3¢ per diaper, including detergent, water, electricity, and depreciated cost of washer and dryer × 7,300 diapers = $219.00)

SOURCE: From *Environmental Action,* March/April, 1989.

NOTE: Typical length of diapering period is 30 months; estimated number of diaper changes per day is 8; total changes per diapering period is 7,300.

[a]Procter & Gamble claims that most parents will pay less for disposables than for diaper services because of discount coupons on disposables and because more cloth diapers will be used due to double-diapering for absorbency.

[b]From *Diapers in the Waste Stream,* Carl Lehrburger, February 1989. Diaper service diapers last longer because less chlorine bleach is used.

[c]From *Consumer Reports,* August 1987.

[d]From the National Association of Diaper Services survey, August 1987.

[e]Average retail cost in January 1989, per Gerber Children's Wear.

[f]From *Diapers in the Waste Stream,* Lehrburger.

burger, whose $15,000 study was funded by the NADS, concluded that diapers make up at least 2 percent of waste by weight. In 1988, this would mean over three million tons of disposables were in the waste stream.

According to Lehrburger, this deluge of dirty diapers is costing us $360 million to dispose of in landfills and incinerators annually. "For every dollar consumers spend on single-use diapers," he calculates, "there's an additional hidden cost of nearly 10 cents for disposal."

In recent years, legislators in Oregon and Washington considered a variety of bills to tax, regulate, or ban the sale of disposables in their states.

As 1989 legislative sessions kick off in state capitals across the nation, diapers are certain to get legislators' attention in a number of other states.

Procter & Gamble thinks diapers are being unfairly singled out. "We're not saying diapers are not a problem," says P & G spokesperson Scott Stewart, "but they're a small part of the problem."

"We disagree," responds Erica Guttman, an environmental planner with the Rhode Island Solid Waste Management Corporation. "Not only are disposable diapers in a large and growing portion of the waste stream, but more important, they're a perfect illustration of what's wrong with the mentality that disposable is better."

At 2 percent, diapers are the largest single product in the waste stream after newspapers (6.8 percent) and beverage containers (5.5 percent). Recycling experts point out, however, that the diaper share is set to expand as more states implement mandatory recycling programs that pull newspapers, beverage containers, and other recyclables out of the disposal heap.

Even more controversial than the volume of discarded diapers is the question of whether dirty diapers in landfills pose a threat to public health.

Although labels on some brands of disposables recommend emptying baby's excrement in the toilet, Lehrburger estimates that less than 5 percent of parents actually rinse out single-use diapers. The result, he finds, is that three million tons of untreated feces and urine end up in landfills rather than in the sewage system every year.

More than 100 different intestinal viruses are known to be excreted in human feces, including polio and hepatitis. A 1975 study by Dr. Mizdra Peterson published in the *American Journal of Public Health* found that viruses could survive up to two weeks in a landfill.

On the other hand, Procter & Gamble says such viruses are killed by landfill leachate. And a study conducted by Washington State's infectious waste program, says program director Wayne Turnberg, found no evidence of groundwater contamination due to diapers, nor signs that sanitation workers are at risk from handling them. "I wouldn't go so far as to say there's not a problem," Turnberg adds. "We just haven't seen the evidence."

While using reusable cloth diapers is clearly the most environmentally sound choice in baby-diapering, there are substantial barriers to increased use of cloth.

"One of the problems is that 75 to 85 percent of new mothers are automatically introduced to disposables in the hospital when their baby is born," explains Pat Greenstreet of the King County Nursing Association in Seattle, Washington. The nurses' group, which every year identifies three health issues to work on, received a grant from the city's engineering department to educate parents and child-care professionals about the diaper dilemma.

The second institution in many childhoods—the day-care center—often sets up another roadblock. Most day-care centers require that parents provide a day's supply of diapers, and many allow only disposables.

Margot Guthrie, district manager of Kindercare in Tampa, Florida, explained that her company requires disposables to prevent the spread of infection. "They're easier, more sanitary, and have fewer health risks," she says. At Kindercare, parents wishing to provide cloth diapers need a doctor's note saying a cloth diaper is required.

Paul Fogel, president of DyDee Diaper Service in Tampa, says his company has mounted a successful campaign to overcome resistance from day-care centers. According to Fogel, some centers in his area had discouraged cloth diapers by misinforming their clients that Florida state law prohibits use of cloth diapers in day-care centers. Fogel's company sent a general letter asking day-care centers to refrain from that practice. DyDee Diaper has also begun sending customers a list of day-care centers that allow cloth diapers along with the usual batch of health and child-care information for new parents.

Day-care centers can also subscribe directly to a cloth diaper service and include the cost of the service in their clients' bill. Fogel's company, for example, services about 30 day-care centers in five counties.

Currently, however, diaper services are not available to everyone. Some barriers are geographic — most diaper services are located in urban areas. Another barrier is public assistance programs that discourage use of diaper services. For example, the federal WIC (Women, Infants, and Children) program permits the purchase of disposable diapers but does not include cotton diapers or diaper services on its list of purchasable products. Another economic disincentive to the wider use of cloth diapers is import quotas recently imposed on cotton by the federal government, which threaten to increase the cost of cotton diapers by as much as 25 percent.

Despite signs of changing attitudes, the most significant hurdle for cloth diapers is the current widespread consumer preference for disposables.

Tina Barry of Kimberly-Clark in Cincinnati argues that "the popularity of our diapers has grown in direct proportion to product improvements." These include introduction of elasticized leg bands in the early 1970s, extra-absorbent materials later in the decade, and then reusable tabs.

"These things are very good products," says Kay Willis, founder of Mothers Matter, an organization that provides educational seminars on parenting from its Rutherford, New Jersey, base. "I raised my ten kids on cloth diapers. My grandkids all had disposables. There's just no comparison." Willis argues that disposables are more comfortable for babies, allowing them to sleep through the night and reducing the incidence of diaper rash.

Cloth diapers have their own loyalists. "Laundering diapers at home is the least expensive method of diapering," writes pediatrician Rebecca Stauffer in an Arcata, California, co-op newsletter. Stauffer also cites environmental and health benefits of cloth diapers. "As a working mother, I do not find this an offensive or burdensome choice."

When General Health Care Corp. surveyed its Philadelphia-area customers, 92 percent said they were pleased with the company's diaper service, the *Philadelphia Tribune* reported. Asked why they subscribed, 31 percent said cotton diapers were better for a baby (with 23 percent saying cotton diapers don't cause rashes) and 26 percent said they were better for the environment.

Not surprisingly, both combatants in the diaper war claim their products reduce diaper rash — on one side because diaper service diapers are pH controlled and rinsed in a bacteriostatic germ-killing agent, and on the other because super-absorbent throwaways reduce skin wetness and the mixing of urine and feces.

Diaper services claim that the American parent has been blinded so far by "the seemingly inexhaustible advertising dollars of the disposable diaper industry," as Jack Mogavero, president of General Health Care Corp. puts it.

Diaper services—turned environmentalists—are fighting back with some aggressive advertising of their own. "The argument for disposables is a lot of garbage," reads a NADS ad in which a full-to-the-brim "garbage barge" heads out to sea.

In a separate public relations campaign, General Health Care Corp. is sending "Crusader Baby" to malls and other public spaces to educate parents and child-care professionals about diaper services.

Whether the quality differential is real or merely perceived, disposable diapers will prob-

---

Earth Care Action

Diaper Solutions
By Jeanne Wirka

To solve the diaper waste problem, Environmental Action's Solid Waste Alternatives Project recommends that individuals, governments, and industry take a number of steps. These measures conform with the following hierarchy (with best coming first) of solid waste disposal options: Reduce the amount of waste produced, reuse, recycle, or compost, and—least preferable—incinerate or landfill.

Individual Actions

● Choose cloth diapers. Consider using a diaper service if one is available in your area.

● If your child attends day care, look for a center that allows cloth diapers. If your current day-care center doesn't, ask for a change in policy to allow parents to bring cloth diapers or to subscribe directly to a service.

● If you do use disposables, take care to flush diaper contents before disposing of the diaper. Also, write to the disposable diaper makers and encourage them to develop a flushable liner.

Federal and State Government Actions

● Require that cotton diapers be used in all government-funded or operated institutions (such as hospitals and child-care centers).

● Modify government assistance programs such as the federal WIC (Women, Infants, and Children) program so they do not discriminate against cotton diapers.

● Provide consumers with information about diapering alternatives through appropriate health and environmental agencies, and by providing grants to local citizens groups or professional associations for education campaigns on reducing diaper waste.

● Provide tax incentives for setting up new diaper services.

● Conduct thorough investigations of the potential impacts on public health of landfilling disposable diapers.

● Levy a "disposal fee" on manufacturers of disposable diapers.

● Encourage the development of environmentally safe composting systems for diapers and other waste through technical assistance and government financing programs, research into methods to remove toxic components of municipal waste and sewage sludge, and government procurement policies favoring compost products.

Industry Actions

● Develop and market a disposable diaper with a flushable lining.

● Support research into composting techniques for diapers.

● Encourage customers to flush diaper contents.

From Environmental Action, *March/April 1989.*

ably be with us for some time. Accordingly, some waste experts are looking for better disposal options.

The flushable diaper. Lehrburger thinks that a flushable, single-use diaper is both technically feasible and desirable. He envisions a two-piece diaper, with a reusable diaper cover and a highly absorbent flushable inner padding. The flushable diaper, he says, could make today's throwaway obsolete.

Less waste would go to landfills, since the flushable diaper would rely on the normal sewage-treatment system for disposal. And the flushable diaper might offer benefits over cotton diapers, Lehrburger suggests, by eliminating the detergent, energy, and labor involved in laundering.

Compost those diapers away. Composting—the natural decomposition of organic materials by microorganisms—is gaining acceptance in the United States. "Co-composting," which refers to the process of composting solid wastes and sewage sludge together, could be used for disposable diapers, according to Lehrburger. Pulp, paper, and feces would degrade into humus, which can be used as topsoil or soil-enhancement material. The diaper's inorganic plastic parts, however, would have to be sifted out.

Biodegradables. One way to make diapers more compostable is to design them with biodegradable plastic shells. At least one company has done so. Last September, the Rocky Mountain Medical Corp. announced its "TenderCare" diaper, which utilizes a cornstarch-based biodegradable plastic outer diaper backing.

Many waste officials and environmentalists are skeptical of the new degradable plastics, arguing that little is known about their rates of degradation or what they degrade into. At best, biodegradables are a mitigating strategy, says Lehrburger, since they don't reduce the costs of transporting waste or the cost of landfilling.

Given all this environmental consciousness around diapering practices, some waste-conscious parents have wondered whether laundering cloth diapers—either by parents at home or by diaper services—isn't itself an environmental problem.

General Health Care's Jack Mogavero says that local sanitary districts monitor wastewater from diaper services, and that wastewater must conform to standards for phosphate and chlorine content.

Diaper services are "not a problem to the local sewage system," adds Jack Schiffert of the diaper service association. The waste sent into city sewage systems is "basically the same as household wastewater," he says, but since the volumes are greater, diaper laundry services pay a "sewage surcharge," which varies according to local rates and the volume of effluent.

What's missing is a comprehensive study comparing the environmental impact of washing diapers vs. disposables. Among the factors that could be studied: how wastewater is treated by individual sewage plants, the "front-end" pollution costs of manufacturing plastic and absorbent chemicals for disposables, pollution from the cotton-growing and manufacturing process, and the transportation pollution created by trips to stores to buy disposables vs. diaper service trucking.

This is the essence of the diaper dilemma—the true costs of diapering go beyond those paid directly by moms and dads, diaper services, and disposable diaper makers. Those costs, like dirty diapers, don't disappear.

Reprinted by permission from Environmental Action *magazine, March/April 1989.*

SOLID WASTE

Getting into a Heap of Trouble

By Bill Lawren

Sorting Out the Truth 98
Legislative Avalanche: 2,000 Solid Waste Bills 100
One-Year Moratorium for New Incinerators in Massachusetts 101
Give Us Your Tires, Your Coors Cans... 102
The Consummate Recycler 106
From Landfill to Landscape 110
Landfill Alternatives 111
A Pay-by-Bag Town Thinks Trash 113
Corporations Learn to Turn Their Trash into Cash 115
Recycling Pioneers Are Seeking Ways to Exchange New Life for Old Tires 117
Ontario Recycling in Blue Boxes 118
Once a Month, an Unruly City Scrapes Itself Clean 120

Early in September 1986, the barge *Khian Sea* set out from Philadelphia to begin what would become a woefully symbolic odyssey. Her load: 15,000 tons of toxic ash, the noxious by-product of Philadelphia's massive garbage incinerators—a cargo, in other words, of the trashman's trash. Over the next 18 months, she was turned away by ports along the American Atlantic, in the Caribbean, even as far away as West Africa. Finally, last February [1988], she gave up and returned to Philadelphia, her holds still laden with a cargo that no one wants.

Meanwhile, in a grimy, semi-industrial neighborhood in South Philadelphia, a glistening new factory nears completion. This ultramodern plant, full of high-tech sorters and shredders, may someday represent part of a solution to the city's staggering, 6,000-ton-per-day garbage problem. For this factory, built by the Swiss-based ORFA Corp., can take raw solid waste straight from garbage trucks, sort it, chew it up, bag it, and spit it out at the other end as high-quality cellulose fiber or granulated "sand." These products can then be sold to industry, the last step in an almost alchemical transformation that turns garbage into gold.

From National Wildlife, *August/September 1988. Reprinted by permission.*

The *Khian Sea* and the ORFA factory are potent symbols of a struggle that has embroiled not only Philadelphia but countless communities all across the country. The problem, simply stated, is what to do with the mounting heaps of trash we Americans are generating. According to Neil Seldman of the Washington, D.C.-based Institute for Local Self-Reliance, a nonprofit group that advises municipal governments on garbage problems, Americans now generate some 200 million tons of solid waste a year, more than twice as much as in 1960. Yet the traditional response—bury it—obviously is no longer adequate; landfills are closing in this country at a rate that averages out to about one per day. As a result, cities are scrambling to find places to put refuse.

San Francisco has been shipping its orphan garbage to out-of-town landfills for 20 years, and is now paying $2.5 million just to *reserve* future dumping space in Solano County, some 40 miles away. Beleaguered Philadelphia, which closed its major landfill just across the Delaware River in New Jersey in 1984, now trucks some of its garbage to Dundalk, Maryland—more than 100 miles away! "Within three years," says Seldman, "more than half our cities will exhaust their current landfills. The emergency is real."

As the garbage crisis heats up, many communities find themselves in the throes of a growing acrimonious political struggle. Last year, for example, five out of seven candidates for the New Jersey state legislature identified waste management as the number one issue in their campaigns, while in Nebraska, former Lincoln Mayor Roland Lutke told the *Wall Street Journal* that "garbage is the kind of issue that can unseat an incumbent mayor."

In Philadelphia, where Mayor W. Wilson Goode has supported the building of a new $280-million garbage incinerator, protesters have appeared at public hearings carrying signs that bear such slogans as "Goode's Taking Our Breath Away." In San Diego, citizens angry about garbage problems waved dead rats in the faces of officials; one six-year-old girl asked former California Waste Management Board Chairman Terry Trumbull, "Why are you trying to kill me?"

Finger pointing and rat waving still characterize the garbage battle in many communities. But even in those places where the clatter of political conflict has been loudest, everyone

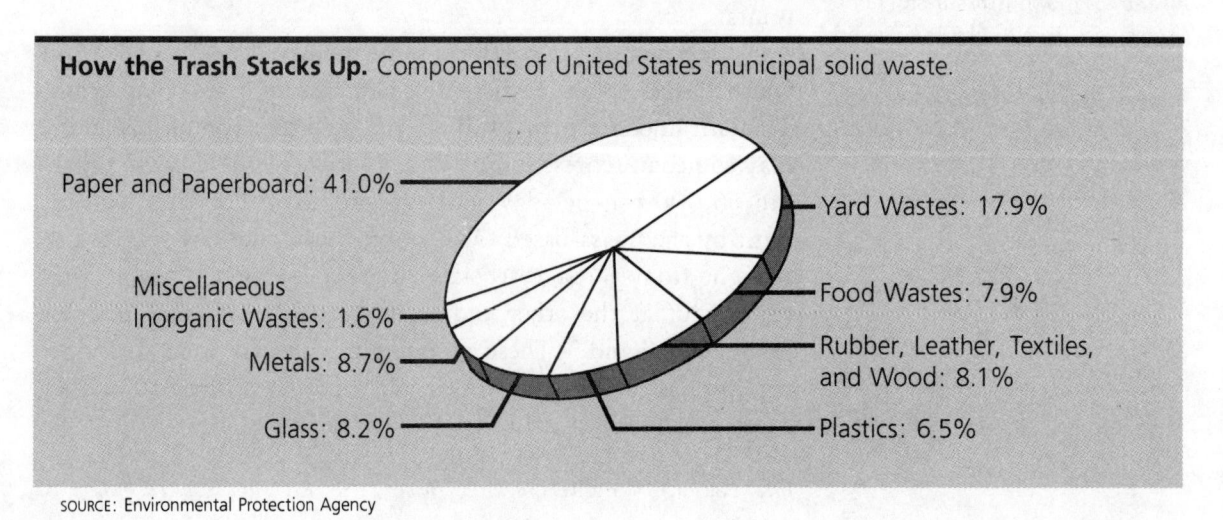

How the Trash Stacks Up. Components of United States municipal solid waste.

- Paper and Paperboard: 41.0%
- Miscellaneous Inorganic Wastes: 1.6%
- Metals: 8.7%
- Glass: 8.2%
- Yard Wastes: 17.9%
- Food Wastes: 7.9%
- Rubber, Leather, Textiles, and Wood: 8.1%
- Plastics: 6.5%

SOURCE: Environmental Protection Agency

agrees that the problem can no longer be ignored or left as a legacy for the next generation. With that fundamental agreement in place, America's cities and towns are beginning to slowly grind out a series of responses to the garbage problem—responses that, though they may in themselves be the source of renewed debate, are still more positive and purposeful than the apathy and neglect of the past. The action is taking place primarily on three fronts.

Landfills

About 90 percent of the more than 500,000 tons of trash generated every day in this country is still buried in landfills. "We'll always need them," says Seldman, "if for nothing else than to get rid of dirt and certain metals and plastics" that can't be disposed of in any other way.

Since the passage of the federal Resource Conservation Recovery Act in 1976, states have been mandated to develop safer waste-disposal plans. As a result, most states now require newly built landfills to comply with strict design standards and construction procedures—including two to three layers of underlinings and regular monitoring of nearby surface and groundwater.

Despite such precautions, however, severe problems remain. Of the approximately 9,500 landfills still operating in this country, Seldman estimates that only about 10 percent are properly permitted. No one knows for sure how many of the others are leaking potential contaminants into the soil and underground water supplies. Authorities do know though, that even if they can afford the escalating costs of creating new landfills, they must still find locations for such facilities.

These days, most people are violently opposed to having a landfill built in their "backyards." Franklin County, Ohio, Commissioner Jack Foulk discovered that fact firsthand not long ago, when he received an anonymous midnight call. "You S.O.B.," the caller reportedly growled. "If you put in a landfill next to us, we're going to kill you." (Death threats notwithstanding, the landfill was eventually built and has since been cited by a national waste-management association as one of the best in the country.)

Incinerators

When burying garbage becomes unfeasible, many authorities think the next best alternative is to burn it. The reasons are clear: Incineration requires only small changes in garbage collection systems, and at $50 to $100 a ton, it can still be cheaper than many landfills. What's more, two newer breeds of incinerators—"Trash-to-Energy" and "Refuse-Derived Fuel (RDF)" plants—can further offset costs by producing salable energy in the form of electricity or steam.

All this has proved attractive enough to make incineration the solution of choice in as many as 110 locales, including Detroit and Westchester County, New York. At this writing [summer 1988], an additional 210 incinerators are either under construction or in planning stages. "Unfortunately," notes Norman Dean, director of the National Wildlife Federation's Environmental Quality Division, "by building new incinerators, we're simply giving people more incentive to create more solid waste. Communities need to be developing programs to *reduce* the amount of waste to be disposed of."

For every municipality that sees burning as the wave of the future, another has come to see it as a Faustian bargain that in the end only creates more problems. For one thing, the building of new incinerators requires a huge capital outlay; in New York, for example, a proposed incinerator to be built in the

Brooklyn Navy Yard is expected to cost as much as $400 million. In addition, a survey by *Editorial Research Reports* indicates that Trash-to-Energy and RDF incinerators, which can potentially offset some of that outlay by producing energy, often have been plagued by mechanical problems.

Perhaps more important to citizens' groups and environmentalists, incinerators may represent a widespread threat to public health. "Because of gaps in the Clean Air Act, cities may operate waste-burning facilities without adequately controlling toxic substances produced in the combustion process," says Dean.

Such incinerators have been shown to emit a worrisome variety of dangerous substances, including acid gases, dioxins, and nitrous oxide. In fact, Eric Goldstein, an attorney at the Natural Resources Defense Council, maintains that New York's proposed Brooklyn incinerator would become "the single largest new source of air pollution in the city."

The Environmental Protection Agency has proposed stringent new regulations to help control incinerator emissions, but those regulations will not take effect until 1991. By then, dozens of new facilities may have been built without the extra pollution-control equipment.

Meanwhile, incinerator operators must cope with another dilemma: The ash created by the burning process is hazardous because of its heavy metal content and thus presents a special disposal problem of its own, as the hapless travels of Philadelphia's *Khian Sea* illustrate.

Incinerator proponents maintain that present-day antipollution devices can reduce toxic emissions by as much as 99 percent.

Sorting Out the Truth
By Bill Lawren

To most people, garbage is either a nonentity or a downright nuisance. But to archaeologist William Rathje of the University of Arizona, trash is truth.

"Archaeologists study ancient garbage to learn about past civilizations," says Rathje, who calls himself a "garbologist." "Here we look at our own refuse to study our own civilization. We can learn a lot about our society just by looking at the back end."

That's precisely what Rathje does. Over the past 15 years, he and an army of students have diligently sorted and classified tons of household garbage from cities as diverse as Tucson, Milwaukee, and Mexico City. (No, they don't go door-to-door. Municipal governments collect the garbage for them.)

The results are often surprising. For example, Rathje has found that among the largest components of municipal garbage are newspapers (12 percent) and phone books; fast-food packaging, at one-quarter of 1 percent, is a smaller component. (Most fast-food packaging, of course, doesn't enter a waste stream in household garbage.) The researcher has also found that low-income Hispanic families take more vitamins and use more cleaning products than higher-income Hispanics or Anglos.

Want to know what people really eat? Take a look at their trash, says Rathje. For instance, when fatty meats became a national health issue in the mid-1980s, he discovered that the amount of fat trimmings in garbage almost doubled. However, that was counterbalanced by evidence of increased consumption of ground meat and sausages—meats with "hidden fat."

Almost all of us, he observes, tend to think we eat better than we actually do. "People typically overreport consumption of 'healthful' food like vegetables," he says, "and underreport 'bad' foods like alcohol and fatty meats."

The research is not without its problems. "Dogs and cats love to eat our data," says Rathje. And the odors? "Well," he chuckles, "they do tend to separate the men from the boys." His most recent program, excavation of three municipal landfills, was partially designed to resolve what is perhaps his most irksome problem: lack of regard from other archaeologists. "In the past," he says "they didn't respect us because we didn't actually get down in the dirt and dig."

From National Wildlife, *August/September 1988.*

"Our new plants have all passed dioxin tests with flying colors," says Amy Hildabidle of the Ogden-Martin Corp., which has recently built six new waste-to-energy facilities in various western localities.

Even so, the outcry against incinerators has made many communities nervous. In New York, lawsuits by environmentalists have postponed the building of the Brooklyn incinerator until at least 1992. Three proposed trash-to-energy plants in California have been shelved, as have facilities in Lowell, Massachusetts, and Broward County, Florida.

Recycling

To many people, the most demonstrably commonsense approach to getting out the garbage is to make less of it in the first place. Neil Seldman estimates that by employing several recycling tactics, Americans might eliminate as much as 80 percent of their waste stream.

The process, he says, should start in the home, where as much as 40 percent of a city's garbage can be separated and subsequently reused if a strong market exists. Since 1972, when Oregon passed a bottle recycling law, some state and municipal governments have seen the benefits of such programs. (Eight states and one city, Columbia, Missouri, now have bottle bills.) New York's four-year-old bottle law, which charges 5-cent deposits, has resulted in return rates of as much as 80 percent.

Bottles, though, are only the beginning. The Philadelphia City Council recently passed a law that established a goal of a 50 percent recycling rate by 1991. Three states—New Jersey, Rhode Island, and Oregon—have enacted bills that now mandate statewide recycling. (The New Jersey act, which was passed last year, requires every state resident to participate.)

In some places, home recycling, or "source separation," is encouraged by incentives, both positive and negative, that sometimes border on the ingenious: Islip, New York, employs a "garbage girl" who checks trash cans and levies fines for repeated noncompliance; authorities in Rockford, Illinois, award a $1,000 cash prize each week to "good-guy" recyclers.

As source separation takes hold at the household level, industry is gradually getting into the recycling act. Commercial giants like Eastman Kodak and small companies like St. Jude Polymers in Frackville, Pennsylvania, are manufacturing recyclable plastics for use in making containers, upholstery fill, and even clothing. Rubber Elastomers of Babbit, Minnesota, has developed a process by which old tires can be crumbled to make an underpinning material for walls and floors.

But for all its apparent appeal, for all the ardor it generates in its supporters, recycling continues to lag in this country. Despite ambitious goals and education programs, America still recycles only about 10 percent of its trash. Traditionally, many industry officials have been reluctant to use recycled raw materials—on the one hand, there have been concerns about uniform quality and dependable delivery of those materials; on the other, there have been considerable tax advantages on the books pertaining to the use of virgin materials. The situation is compounded by the fact that demand for recycled products has yet to match supply.

As the dust of debate settles, many communities will have to attack their growing garbage piles with a multifaceted strategy. In Philadelphia, the Institute for Local Self-Reliance has recommended just such a combined plan: Close both of the city's incinerators, cancel the building of a third incinerator, implement a mandatory home recycling program, turn food and yard waste into compost for use in agriculture and city parks, and for the long term, contract as much disposal as possible to a number of ORFA-style recycling plants.

A similar, though not quite as varied, strategy is already on-line in Marion County, Oregon. There, an aggressive recycling program mandated by the state removes 95 percent of soft-drink and beer containers from the waste stream, as well as thousands of tons of cardboard, newspaper, aluminum, glass, and scrap metal. This refuse is picked up by private haulers and sold; citizens are rewarded in the form of reduced trash-collection charges.

Four years ago, the county contracted Ogden Martin to build a $47-million incinerator with a state-of-the-art emission control system that is now viewed by some authorities as a national bellwether. Preliminary recycling of batteries has greatly reduced the amount of heavy metals in the incinerator ash; the ash itself is buried in small, dedicated "monofills" with two bottom liners and a system that collects and tests the leachates. So far, according to County Commissioner Ronald Franke, the leachate has proved clean enough that the county plans to use it to irrigate farmland.

While these kinds of synergistic approaches will go a long way toward alleviating the garbage crunch, the fact remains that America will never solve its solid waste dilemma without developing programs that deal with the root of the problem: the amount of waste generated.

Currently, the United States generates more than twice as much garbage per person as some other industrialized nations. "This indicates two things," says Norm Dean. "First, that it is possible to maintain a high living standard and still produce less waste. Second, that there clearly is room for dramatic improvement in reducing waste generation in this country."

To achieve that end, some states have initiated waste-reduction programs in recent years. In North Carolina, for instance, where authorities now offer incentives to companies that find ways to reduce waste and pollution, 55 of the firms in the program saved a total of $14 million in one year alone. Unfortunately, notes one environmental consultant, "It's not so much that waste reduction is not happening [in the United States], but that it is not happening nearly to the extent it could or should be."

Clearly, as the country's population grows, getting rid of wastes will become a more urgent concern. The question is, are we willing to make the difficult decisions that require balancing present-day realities with a vision of an ideal future?

Earth Care Action

Legislative Avalanche: 2,000 Solid Waste Bills

It comes as a surprise to no one that 1988 was a year of movement in state legislatures for recycling issues. While most of the focus was on states such as Pennsylvania, Maryland, and Florida, which mandated some form of recycling, numerous issues were addressed and passed into law.

According to Ann Matheis at the American

Paper Institute, approximately 2,000 solid waste bills were introduced in state legislatures in 1988, with approximately 850 addressing infectious waste, recycling, litter control, and incineration. Another 300 dealt with packaging and container legislation, including 77 that concerned biodegradability issues, 11 that addressed paper packaging, and 206 for plastics and plastic packaging.

Perhaps the most comprehensive recycling packages were passed by Florida and Pennsylvania. These two states took a broad-brushed approach to solid waste management that included mandating recycling as well as general improvements to their solid waste management programs.

One tactic the Florida legislation uses to preserve the state's disposal capacity is restricting the options of certain materials. No fewer than five materials—lead-acid batteries, motor oil, yard waste, white goods, and tires—are banned from disposal facilities.

Pennsylvania's legislation directs that the state's 400-plus most heavily populated municipalities develop recycling programs, with the state picking up at least 90 percent of the startup costs. In one of the legislation's market-development provisions, the State Department of General Services will implement a preference program for the purchase of products made with recycled materials. It is also directed to adhere to the federal guidelines on procurement being developed by the Environmental Protection Agency.

Summarizing even the most basic elements of the legislation passed in 1988 is the subject for a book, not a magazine article. However, some of the more noteworthy provisions from this avalanche of legislative activity follow.

- California Assembly Bill 3299 requires that all rigid plastic containers sold in that state must have their composition identified.

- Connecticut House Bill 5885 requires the State's Commissioner of Public Works

---

Earth Care Action

One-Year Moratorium for New Incinerators in Massachusetts

Last December [1988], Massachusetts instituted a one-year moratorium on construction of new burn plants. According to the commissioner of the state's Department of Environmental Quality Engineering, the state issued the ban because enough refuse is currently burned to meet the 50 percent incineration level specified in its solid waste plan.

The moratorium affects planned facilities that do not have a draft permit for operation. It will not close the six existing facilities or the two with draft permits. As specified by the state's Solid Waste Act, 25 percent of solid waste is to be recycled, 25 percent landfilled, and 50 percent incinerated. According to department figures, only 8 to 10 percent is now recycled, more than 50 percent is incinerated, and the balance is landfilled.

Included in a ten-point solid waste action plan is the following: 10 percent waste reduction by 2000; public and private materials recovery facilities such as those planned for Springfield, Boston, and Cambridge; 100 percent leaf composting by 1992; regional initiatives to establish compost facilities (such as in the Berkshire regions of western Massachusetts); and implementation of a solid waste master plan by November 1989.

From BioCycle, February 1989.

to consider recycling when altering any real assets. The bill also requires the Commissioner of Economic Development to adopt a plan to promote industries that use recycled materials.

- Iowa House Bill 2453 provides for tax incentives for the use of degradable packaging products.

- Illinois Senate Bill 1616 requires Chicago and the state's 17 largest counties to develop comprehensive waste management plans which, among other things, ensure that 25 percent of the municipal solid waste is recycled within five years.

- Rhode Island Senate Bill 3050 creates a permanent commission to produce a plan for recycling plastic products and develop guidelines for the use of photodegradable and biodegradable products.

- Washington Senate Bill 6446 requires the Department of General Administration to develop a directory of businesses supplying products containing recovered materials.

From BioCycle, *February 1989. Reprinted by permission.*

Earth Care Action

Give Us Your Tires, Your Coors Cans...

By Talbot Bielefeldt

Our guide has a problem. Dan Geyer's big raft, loaded with camping equipment, is perched on a rock at the head of a rapid, Oregon's Rogue River pouring off on either side. It's an embarrassing place to be stranded, with 11 customers, two reporters, and his lead guide watching from the eddy below.

The tires don't help. Truck tires, car tires, whole ones and chunks of tread—the raft is piled high and weighted down with old rubber. But Geyer can't jettison a one. The tires are our reason for being there.

The Rogue River begins near Crater Lake, drops out of the Cascades and flows past the cities of Medford and Grants Pass, then cuts a deep gorge through the Coast Range forests to the Pacific Ocean. In 1968, Congress determined that 34 miles of that gorge merited inclusion in the national Wild and Scenic River System. The stretch was designated "wild," a classification that would protect it from dams and other federal projects. The Rogue offers excellent scenery, boating rapids, salmon and steelhead fishing, and possibly more abandoned tires than any wild river in the country.

In 1983, the Sierra Club and Oregon River Experiences (ORE), a whitewater outfitter based in Junction City, Oregon, began cosponsoring cleanup trips down the Rogue, giving river enthusiasts the opportunity to bag litter and remove tires. Although the club canceled its river program in 1986 for insurance reasons, individual club members continue to sign up directly with ORE. This September afternoon [1988], they make up most of the audience watching Geyer struggle with his raft.

Unable to move the boat against the force of the river, the guide takes the opposite approach, leaning on the pontoon closest to the water, giving the current even more to work with. The raft pivots slowly on the rock, begins to slide, and then plunges over the edge. Geyer floats down to us with an embarrassed grin.

No one is about to criticize; we have all had our own relationships with rocks. Like most ORE trips, this one is "row your own," with two people to a raft. The guides coach and stand by for rescue, but in the rapids the boats are all ours. On cleanup trips the mental concentration required in whitewater is intensified because we're constantly on the lookout for trash.

Policing the river starts out as altruism but quickly develops into a sport of its own. Tires are the big game. Cans are of intermediate status, the more the better. Cigarette butts are "trash fish." We find fertile hunting grounds just off the river in the oak and madrone along Whiskey Creek, where an old miner's cabin is preserved along with antique mining implements and very contemporary piles of aluminum beer cans.

The cabin, however quaint, and its surrounding litter serve as a reminder that wild and scenic rivers outside Alaska are rarely true wilderness. Maine's Allagash flows through old logging country and is interrupted by dams. The Salmon River in Idaho begins to pick up tires and other urban litter as soon as it leaves the protection of the wilderness areas around its upper stretches. Popular whitewater streams outside the national system suffer greater indignities. In West Virginia, boaters contend with flotsam the New River dumps over the Bluestone Dam, while power-plant releases make the water in the late-summer Cheat River uncomfortably hot for miles below a popular put-in.

Miners and boaters have been part of the Rogue since the middle of the last century. A well-worn backpacking trail runs along the north bank, roads carve close to the scenic corridor of the river, and several permanent lodges are served by boat and small aircraft. Many of the river's rapids exist in their present form because pioneer guide Glen Wooldridge dynamited boulders out of the way.

The Rogue today continues to be an accessible and popular stream, and management focuses on controlling recreational use. In 1974, the U.S. Forest Service, the Bureau of Land Management (BLM), the state Scenic Waterways Commission, and other concerned agencies began limiting the number of boaters. The present system allows 120 people per day to enter the Wild and Scenic section during peak season (between June and early September). Of those spaces, 60 are allotted to clients of commercial outfitters and 60 are open to the general public through a Forest Service lottery.

The procedure is more or less typical of controls on U.S. rivers under federal jurisdiction that boast wild stretches long enough to support multiday trips. The most famous permit system—or infamous, depending on whom you talk to—is for the Grand Canyon of the Colorado, which has a years-long waiting list for private parties. Many boaters maintain that the National Park Service's division of permits is unfair, since private boaters often can't get on the streams while commercial guides may have spaces to fill. On the Rogue, the Forest Service and the BLM have tried to minimize the conflict by putting unused commercial spaces into a common pool, from which private boaters can claim them on a first-come, first-served basis.

Although the absolute numbers and method of administration are open to debate, guides, private boaters, and federal agencies agree in general on the need for some limits on use. River managers on the Rogue point to September days after permit restrictions are lifted when more than 400 people have started down the river. The result has been crowds of "giggle-and-splash" rafters competing for space with anglers trying to quietly drift flies past steelhead trout. In response to the fall rush, the managing agencies extended the end of the permit season in 1986 from Labor Day to September 15. The move was intended to keep the open season free of the holiday crowds, moving it closer to Oregon's winter rainy season.

As far as we can tell, the agencies succeeded. On September 23, our third day on the river, the morning fog thickens into a cold drizzle. Ridge silhouettes vault into the clouds, and spiky trees float on a gray mist. We hear rapids long before we can see them. Hunkering down over our oars, not stopping to scout, we follow the guide boats through one riffle after another. For most of the day we see only the Rogue's native residents. Herons and osprey cruise the canyon, while mergansers paddle back and forth in the eddies, watched by dark-eyed otters.

At Winkle Bar, where Zane Grey lived and wrote in the 1920s, several of us ferry across the river to clean up a camp area on the opposite bank. A guided party of fishermen in cowboy hats eats lunch around a fire. We wander the perimeter of their site, dismantling fire rings and collecting litter. We're self-conscious, aware of certain stereotypes: Fishermen in cowboy hats leave litter and spoil the wilderness experience; we rafters talk environment, but actually crowd the river and spoil the fishing. In fact, our party includes several anglers, and fishing guides on the Rogue and most other regulated rivers operate under the same rules as everyone else does—fires only in metal fire pans, all refuse and human waste packed out.

According to the river's managers, the trait shared by individuals who soap up in streams, defecate on beaches, and build fire rings in camp is not river-running savvy but ignorance. Fire rings on the Rogue appear mainly after permit season, when boaters are not required to check in with the Forest Service and don't get the official pitch on fire pans and camping practices. And numbers alone don't make a dirty river. Only 11,000 boaters float the Rogue during the permit season, and the Forest Service estimates that the total number for an entire year is less than 20,000. Some whitewater streams of the East—Pennsylvania's Youghiogheny, West Virginia's Gauley, North Carolina's Nantahala—have seven or eight times that amount. Yet boater-deposited litter is not a major problem on these streams. Most runs there are single-day, so camping is not an issue. And with so many people around, it's hard to chuck a beer can into the water without witnesses.

As a cleanup party, we are generally hypersensitive to litter. But as the hours wear on and the rain soaks in, the real trashbusters emerge from among the fair-weather caretakers. The champion tire-spotter of the group is a Seattle bus driver whose life on wheels seems to have given him a sharp eye for rubber. Late on a drizzly afternoon, while most of us hope only for camp around each bend, he still burns with the spirit of the hunt. "Tire! Right bank! Two on the left! Pull into the eddy!" he shouts.

The day is filled with tires—heavy, cold, wet tires covered with coarse sand that abrades numb fingers. The lead guide, Gregg McKay, grins as we help him pile another hunk of tread onto his sagging boat. "I think we broke the record," he says. Four of the Sierra Club people who had been in eastern Oregon on an Owyhee

River cleanup earlier in the summer are impressed. Their take was not nearly so large. "Over there we were looking for any old cigarette butt we could find," one of them says.

Comparing the impacts of humans on different rivers is tricky because so much depends on factors other than boater behavior. The Owyhee is cleaner than the Rogue partly because it has enough water for whitewater runs during only a short time each year. The Western River Guides Association maintains that the Gunnison River in Colorado owes most of its litter to walk-in fishermen, despite its quarter million yearly rafters. The Arkansas, heavily used by boaters, is comparatively clean. The Green and Colorado rivers in Utah, although regulated by the BLM and worshipped by boaters, get the same kind of debris that shows up in the Rogue and Salmon. The general rule: The greater the access to a river, the more trash you'll find on a river.

Our major cleanup duties end at our third campsite, near Rogue River Ranch and the start of the whitewater of Mule Creek Canyon and Blossom Bar. (Below the rapids, beyond the river's "wild" section, the Forest Service patrols for litter in jet boats.) Before shoving off in the morning, we haul our trash bags and 22 tires up to the ranch pasture for removal by truck. It takes two or three people to roll each of the larger tires up from the riverbank. The rain falls in a solid downpour, and the cold, slippery tires topple and roll back on the rocky slopes.

There's a certain futility to it all; despite its protected status, the wild Rogue is not invulnerable. New tires will wash in next year, and other intrusions from upstream can't be removed by litter bag and raft. The partially completed Elk Creek flood-control dam threatens to raise water temperatures and damage fish runs. Mining in the Rogue River National Forest exposes the stream to cyanide poisoning. And the Rogue is one of the lucky ones. Dams don't threaten its rapids with inundation, as on California's late Stanislaus, or near-total diversion, as on Tennessee's Ocoee.

Still, it's satisfying to look at a cord-size heap of trash and know that a once-wild place has been even partially cleaned up. Most floaters down the Rogue will probably not even notice that the tires and cans are gone. But if they come to expect to find their river clean and free-flowing, they may be less tolerant of trash and poisons and dammed water on this and other streams.

Late in 1988, Congress passed an omnibus river-protection bill that added the upper Rogue and 39 other Oregon river segments to the Wild and Scenic River System. Oregonians backed up the federal legislation with a state measure adding parts of the Rogue and other streams to their own Scenic Waterways System. Ultimately, protection of rivers depends on lawmakers and law enforcers. But it's the work of river advocates, as lobbyists and as tire-rollers, that helps show them the way.

From Sierra *magazine, March/April 1989. Reprinted by permission.*

Earth Care Action

The Consummate Recycler

By Larry Borowsky

Cliff Humphrey took a sledgehammer to his car in 1969. The public dismantling of the 1958 pollution-spewing Rambler (the last car Cliff and his wife, Mary, have owned) showed the depth of Cliff's commitment to a healthier environment.

Other people wouldn't go so far to protect the earth, Humphrey knew, but there was a less traumatic option—recycling.

Humphrey recognized that the small chore of separating recyclables from outgoing trash could give citizens a sense of personal responsibility for the quality of the environment. Thus inspired, people might eliminate all kinds of wasteful habits.

Twenty years ago, few cities paid attention when Cliff Humphrey suggested that newspapers, bottles, and cans could be systematically collected for recycling. These days, communities across the United States have made recycling a high priority and many are hiring Humphrey's services to help launch local recycling programs.

This irony provides Humphrey, now aged 51, with no small measure of amusement, but he's not the "I-told-you-so" type.

In these two decades of surge-and-slough environmental concern, Humphrey has been one of recycling's most vital and dedicated practitioners.

Humphrey's trend-setting, grass roots organization, Ecology Action, set up one of the country's first recycling programs in Berkeley, California. Soon after, Humphrey introduced a pioneering concept—curbside pickup of recyclable trash—to a nearby city. In a salute to his accomplishments and longevity, Humphrey was introduced as the "grandfather of recycling" at the 1988 National Recycling Congress, an annual meeting of recycling activists and administrators.

"He's always been a hero of mine," says Neil Seldman, who is himself a recycling leader.

"He's taught me a lot. The reason Cliff stands out [from other recyclers] is because they are technicians. Cliff has the technical know-how, but he also has a philosophy," says Seldman, director of the Institute for Local Self-Reliance in Washington, D.C.

"Cliff is an innovator," adds San Diego County recycling coordinator Rich Anthony.

In 1968, the Humphreys and another couple founded Ecology Action. An art exhibition on the world's environmental ills—created by artist Mary—brought in the first members.

It was a loose coalition of individuals seeking to teach the public how to conserve resources and avoid polluting the environment. Members carried groceries home in knapsacks, rather than the standard supermarket brown bag. They avoided excessive throwaway packaging and piled leftover food atop a compost heap instead of into the trash can. Old clothes were patched up, envelopes used and reused and re-reused. Cliff and Mary ceremoniously dismembered their car and then members took a 500-mile "survival walk" through California for Earth Day 1970.

Cliff, who had been a highway engineer in

the army and studied ecology at the University of California, began to push recycling.

For guidance, he looked to the emergency recycling efforts of World War II and the commercial "salvage" industry. Profit in the salvage business depended on pursuing the large, clean, easy-to-collect lumps of waste churned out by commercial establishments.

Since Ecology Action wasn't in it for the money but to change individual attitudes toward waste, the group resolved to pursue the more undisciplined garbage produced by ordinary people.

This would be far more problematic. Commercial waste could be collected efficiently because it was generated at relatively few sites. Residential waste was generated in small amounts at every house in town. If it solved this problem, Ecology Action would still have to find markets for the recovered materials.

Cliff Humphrey dismantled his auto to fight pollution. He also pioneered some of the key recycling concepts that today are reducing our waste. (Photo courtesy of Environmental Action.)

And of course, the proposition couldn't work at all unless residents went to the trouble of keeping recyclable items out of their trash.

"Nobody back then believed the people would fuss with their garbage for environmental reasons," Humphrey recalls. But he had faith that the public would cooperate.

And people did. Early in 1970, Ecology Action began collecting newspapers, bottles, and cans every weekend in 55-gallon drums set up in a parking lot. Almost immediately, Berkeley's "dropoff" recycling center touched off a craze. One observer likened the weekly scene to an exuberant county fair. Hippies, little old ladies, society matrons, and businessmen stood shoulder to shoulder with their boxes and bags of refuse while Ecology Action volunteers handled the drums.

Tremendous volumes of material flowed in every week, swallowing up the group's storage capacity. The group hustled to find buyers for the materials and were pleasantly surprised by the cooperative spirit some local industries displayed. One company even loaned Ecology Action a truck and driver once a week.

Citizens in neighboring communities soon began to organize recycling programs of their own; one year later a San Francisco newspaper counted 71 ongoing recyling efforts in the Bay Area.

The Berkeley program won many admirers, but it had its detractors as well. The critics asserted that dropoff recycling produced negligible environmental benefits. Humphrey recalls one study, conducted by Stanford University engineers, which concluded that the amount of gasoline wasted (and air pollution produced) by people making special trips to drop off their recyclables outweighed the energy and resources saved by recycling.

Humphrey freely concedes that dropoff centers aren't the most environmentally beneficial approach to recycling. "The dropoff was a necessary first, inefficient step," he explains. "It was the way that you established the recycling values in the community."

And changing public values was the number one priority.

Rapid success with Ecology Action's dropoff experiment did not fully satisfy Humphrey. He realized that Berkeley's population, a funky mix of intellectuals and hippies, had an unusual predisposition to try out new ideas. Recycling would need a broader constituency. Could it work in a more typical American city?

In the summer of 1970, the Humphreys withdrew from the booming Berkeley recycling scene and moved inland to Modesto, California, a conservative agricultural town of about 50,000 residents. Arriving with no money, Cliff and Mary lived temporarily in a tent pitched in a friend's backyard. With a few loans, donations, and receipts from the sales of Cliff's elementary school textbook *What's Ecology?*, they soon launched a new chapter of Ecology Action.

Humphrey took his case directly to the public, telling the local media that he intended to turn Modesto into a "model of ecological sanity" for the rest of the nation. Members of the community hardly rejoiced over this news.

"You had a lot of suspicion that a long-haired Berkeley radical was importing something weird from the Bay Area," recalls Peggy Mensinger, an early Humphrey supporter who served as the city's mayor from 1979 to 1987.

But Humphrey's high profile in the local papers and on the radio gradually began to convince people that recycling was not a subversive plot but a commonsense approach to managing natural resources.

Within a year, *Look* magazine reported that Modesto had gone "bananas" over recycling. The Boy Scouts, the Modesto YMCA, several church congregations, and various other groups helped round up money, volunteers, and material for Ecology Action's new dropoff recycling center. Even gas stations and bars began saving their beverage containers. The zeal with which Modesto took to recycling surprised Humphrey himself.

"There's no holding this thing back," he told *Look*. "If we make recycling work here, it will work anywhere."

Modesto's enthusiasm convinced Humphrey that the town was ready for the next phase of recycling—curbside pickup. This idea, first conceived back in the Berkeley days, would make recycling more efficient and more convenient for residents. Initially, it sparked a new flurry of negative commentary.

"Everybody told me that it just wouldn't

work," he remembers. "They said if you put stuff on the street, the wind will blow the papers all over, and the kids will pick up the glass bottles and smash them."

As with all his pioneering steps, Humphrey knew he was taking a risk. "To be honest, we didn't know whether the kids were going to smash the bottles or not," he admits today. "But we went ahead."

Ecology Action put the first pilot route into operation in early 1972 with a pair of secondhand trucks, while city officials fidgeted. "I think they figured, 'Well, let him do that for two years or so and then he'll go away, and our public works people can go back to putting the trash where it belongs,' " Humphrey laughs. "But people kept bringing it out to the curb and we kept picking it up."

By the end of the decade, Ecology Action provided curbside service to all of Modesto and the entire surrounding region. It sent out five collection trucks every day, operated a buyback center, and maintained dropoff sites throughout the area. Recycling had become a Modesto institution, a part of the community's daily routine.

More important, it had changed the outlook of the city's residents. "Cliff brought Modesto new perspectives," says Peggy Mensinger. "He created a consciousness here, a force for environmental action."

Humphrey was named "Recycler of the Decade" at the first National Recycling Congress in 1980. Ironically enough, Humphrey was struggling to keep Ecology Action afloat at the time. The nonprofit group survived almost entirely on the resale of the materials it collected, so when market prices dropped sharply in the late 1970s, debts quickly began to accrue.

"At some point everyone rises to their level of incompetence," Humphrey says wryly. "This is where I discovered mine."

Ecology Action pulled through, but the wearying pace burned Humphrey out. Turning the organization's reins over to a successor, he and Mary headed back to the Bay Area in the fall of 1980. No sooner had he "retired" from recycling than a new opportunity came along. San Francisco city supervisors were looking for someone to put together a comprehensive municipal recycling program. The offer was too good to pass up.

But Humphrey's vision of citywide curbside pickup soon ran into an indomitable city machine. "You have to work six or seven months and finally people start showing you where the doors are to the closets," he comments ruefully.

Humphrey eventually secured state money to get the program off the ground, but he didn't stay on the job much longer. He found himself at odds with city planners who wanted to combine recycling with garbage incineration. Humphrey said no thanks.

Today there's a booming demand for people who know how to make recycling work. A critical, nationwide shortage of landfill space and skyrocketing garbage disposal costs have forced communities to reduce the volume of waste that goes to the dump. Currently employed by a private consulting firm in San Francisco, Humphrey spends weeks at a time on the road helping local officials plan and execute recycling projects.

Recycling is only beginning to hit its stride, Humphrey feels. "Right now we recover around 12 percent of the waste stream," he says. "In theory we could do 50 percent. If you compost the organic stream—leaves, yard clippings, and so forth—and recycle the cardboard, paper, and glass and metal containers, you've taken care of almost half of the overall weight."

Local and state governments are investing millions of dollars to develop recycling, which they now regard as a potential savior. Humphrey sees this as a crucial test.

"A lot is being asked of recycling right now," Humphrey cautions. Achieving recycling's potential will require more efficient collection methods and new markets for the resale of the materials collected.

It will also take a massive education campaign to make the public more diligent in its recycling habits. If Humphrey's career has proven anything, however, it is that these obstacles can be overcome.

The programs he built in Modesto and Berkeley, and helped launch in San Francisco, have become lasting successes. Half Berkeley's population recycles at least some garbage today, and the local government has set 50 percent of the waste stream as an official recycling goal to be reached by 1991. The San Francisco program, now in its ninth year, recycles about 240,000 tons of garbage annually, about 25 percent of the city's solid waste output. Modesto's recycling program was taken over by the city in 1982 and later by private contractors.

These developments meet with Humphrey's approval. But he is most pleased by the overall level of public concern over environmental issues. "The newspapers wouldn't give so much space to acid rain and the greenhouse effect if they couldn't take an interested readership for granted," Humphrey says. "I'd like to think that all the years of recycling have helped bring that along."

As in conservative Modesto, recycling has gotten the entire country onto the right path, Humphrey feels, but there is still a lot of ground to cover. He points out that America still needs to burn less gasoline, use less electricity, and stop dumping so many toxics into air and water.

If it has taken 20 years and a solid waste "crisis" for such a painless adjustment as recycling to gain acceptance, will Americans make bigger sacrifices for environmental well-being?

"I know that these ideas can catch on," Humphrey insists. "I've seen it happen with recycling. It will take time. But it can be done."

Reprinted by permission from Environmental Action *magazine, March/April 1989.*

Earth Care Action

From Landfill to Landscape

By Elizabeth A. Brown

It's an eyesore. The gravel-covered, 50-acre mound rising five feet from barren ground behind a shopping center in north Cambridge, Massachusetts, was for 30 years the city landfill.

But if the Massachusetts Department of Environmental Quality Engineering approves the city's plans, today's wasteland will be next summer's [1990] playground—the Thomas W. Danehy Park—complete with three football fields, two softball diamonds, nature walks, and picnic grounds. All for $11 million.

"Many communities look at landfills as derelict lands. But they can be assets," says John Kissida, an environmental engineer at Camp

Dresser & McKee Inc., the Boston consulting firm in charge of the project.

The idea of turning a trash heap into a civic asset—a park, a golf course, a ski mountain, a sports arena (Meadowlands in New Jersey), or a library (John F. Kennedy in Boston)—is catching on. "The national trend is to use [closed landfills] for the public good," says Edward Repa, deputy director of waste programs at the National Solid Waste Management Association, a trade organization in Washington, D.C.

One reason is that landfills are filling up rapidly. Within the next five years, 41 percent of the 6,034 municipal solid waste landfills in the United States will use up their currently permitted capacity, says Allen Geswein of the Environmental Protection Agency's (EPA) municipal solid waste program. Not only will communities have to open new landfills (or find alternatives to solid waste disposal) they will also have to close the old ones properly and plan new uses for them.

"So far, parks are the most popular end-use," says Repa. "With high-density populations and decreasing space for open areas, communities look to recreation.... You never seem to have enough places to play ball."

Because settling garbage creates an unstable base, little else can be done with the space, at least in the first few years. If the landfill is covered with the usual two feet of soil, a large stretch of pavement (basketball and tennis courts, parking lots) will often not remain level. To construct buildings on the site requires complex foundation planning and is not recommended, says engineer Kissida, because of the methane gas created by decaying trash.

Nor is concern over methane limited to the immediate area. "Methane gas gets into utility

Landfill Alternatives
By Elizabeth A. Brown

"When you cap a landfill," says Jim Warner of the Natural Resources Development Council, "it gives you a false sense of security. You can plant stuff on top of it and forget the problem below.

"The point is that potentially valuable land has been preempted by the practice of landfilling," Warner says. "Instead of simply directing our concern to the [environmental safety of] the park, we should question the practice of landfilling and begin recycling."

According to Warner, the United States recycles less than 10 percent of its municipal solid wastes. Roughly 10 percent is incinerated; the rest goes to landfills. (By comparison, Japan recycles up to 75 percent.)

Americans generate approximately 3.5 pounds of trash per person each day. What to do with all this trash?

The Environmental Protection Agency's Agency for Action includes four areas for community attention: reduction (of wastes), recycling, incineration, and landfill.

The Environmental Defense Fund (EDF) suggests reduction and recycling rather than incineration. "Many materials in our everyday waste are toxic and shouldn't be burned," says Richard Dennison, senior scientist at EDF, explaining that lead (found in colored plastics), cadmium, and mercury become more toxic when they are burned. It's dangerous, says Dennison, because this toxic ash is usually added to ordinary, unlined landfills.

Although technological advances in incinerators claim 99 percent pollution-free emissions, Dennison says, this occurs only in optimal conditions, when the equipment—too expensive for many communities—is new.

This year [1989], Seattle began a city-wide recycling program with curbside collection. The recycling goal is 60 percent by 1993. The decision came after cost analysis determined that recycling was cheaper per ton than incineration.

To make recycling successful, says Dennison, government and the private sector must work together to provide markets for the materials. Industry must prepare to change. Incineration and landfill are easy options today because that's the way it has always been done.

"Recycling involves a whole different mentality," Dennison says.

From the Christian Science Monitor, *April 24, 1989.*

pipes and can go into people's basements, build up, and then one spark and it blows up," says Jim Warner of the Natural Resources Defense Council in Washington. The gas can travel as far as a quarter-mile.

Kissida maintains that the Cambridge landfill has been monitored for more than 17 years and has never registered unacceptable levels of methane. The city will continue annual monitoring of methane levels—above- and below ground—after the park is complete.

Another environmental concern is groundwater contamination from leachate—the toxic substance that results when water percolates through waste—especially if the landfill is close to a water source. Although the landfill in Cambridge is only 1,000 feet from the city's reservoir—Fresh Pond—the city manager's office says semi-annual testing has yet to reveal any toxic compounds from the landfill.

Leachate can be contained by a liner within the landfill (modern geotextile materials are sturdy, but expensive). Or if the landfill is built into a naturally nonporous substance such as clay, says Repa, a liner may be unnecessary. The Cambridge landfill was once a clay mine, and dumping began before liners were available. In this case, liners will be used only beneath the football fields and under the meadow marshlands, to protect people and wildlife above from the leachate below.

While some people are concerned about hazardous-waste leachate, Repa says, this is not a problem in a municipal sanitary landfill, where "household hazardous wastes comprise less than one-tenth of 1 percent of the total waste stream." But Warner notes that national requirements for lining landfills and restricting hazardous wastes didn't take effect until 1980. Older landfills, therefore, need more careful monitoring.

Finally, the landfills must be "capped and closed" properly by the landfill operator. Methane must be properly disposed of, and leachate contained. Closure standards are set and monitored by each state, although the EPA recently proposed that more stringent guidelines be enforced nationwide.

Before Cambridge can open bidding on construction of the park, the state environmental engineering office will review the history of the landfill, the measurements of methane levels, and the procedure used for closure.

Kissida is convinced that the methane present in the Cambridge landfill is well below the acceptable level and that the site will pose no threat of contamination to nearby water. To be sure, park plans include a wetlands that will contain excess rainwater as well as serve as a nature preserve. Trees will probably grow, says Kissida, because of the extensive cover already put atop the landfill—4 to 40 feet of mixed materials from tunnel excavations by the Massachusetts Bay Transportation Authority. In addition, 1 foot of loam will cover the entire area.

"It'll be nice to have trees and grass—something to look at besides Zayre [a discount department store]," says Janet Lyman, an area resident, pointing to the mini-mall next to the landfill.

But not all citizens take such a positive view.

"It's just gonna be a place for drug pushers and gang violence," says John Grady, a resident of federally subsidized housing across the railroad tracks from the landfill and shopping mall. He will probably not allow his 10- and 12-year-old sons to play there. Besides, he says, crossing the tracks to get there is too dangerous. "We need a walkway before we need a park."

"I'll be surprised if they ever finish it," says Pamela Cyr, walking her dog past the fenced-off landfill posted with methane-test warning signs.

But Councilwoman Alice Wolf thinks the people of Cambridge are generally satisfied with the way the landfill has been tested and proved safe. "Citizens' groups asked a lot of questions at first," Wolf says. "The city has done whatever it has to to assure the citizens that it will be a safe place."

"The whole process was an example of involvement, agency involvement—federal, state, local," says Kissida.

Expected to cost more than $11 million—$122 per person for 90,000 Cantabrigians—the park will be funded from various sources: $2 million from the state's Division of Conservation Services; $2.9 million from the Massachusetts Bay Transportation Authority as payment for temporary use of the space; $2.5 million from the city, with an additional $4 million to be voted on in May [1989].

Reprinted by permission from the Christian Science Monitor, *April 24, 1989. Copyright © 1989 the Christian Science Publishing Company. All rights reserved.*

Earth Care Action

A Pay-by-Bag Town Thinks Trash

By Andrew Maykuth

In some towns, people greet each other on the street and chat about the weather or the high school basketball team.

In High Bridge, New Jersey, a hilly town of 3,662 near the Pennsylvania border, they talk trash.

"Whenever two people from High Bridge get together, we talk about garbage," said Emily Bruton. "It's always on our minds. People outside of town think we're fanatics."

Solid waste became a household concern last year after High Bridge officials devised a scheme to drive home the soaring cost of trash disposal. They began making residents pay for garbage by the bag. This year, residents paid a flat fee of $140 for the right to dispose of one 30-gallon bag of trash each week; additional bags cost $1.25.

Faced with a charge based on the amount of waste they generated, residents responded with imaginative ways to reduce their trash.

Some stomped on their garbage to get it into one container. Others changed their buying habits after they began thinking about product packaging. "I stopped buying soda in plastic bottles because the bottles take up so much room in the trash," Bruton said.

Recycling became so popular that the civic organizations that collected glass, cans, and paper were inundated with new material. The volume became so unwieldy that the borough took over collections in September [1988].

The bottom line is that after 12 months, this Hunterdon County borough has reduced the amount of garbage it collects by 25 percent.

High Bridge's experience is similar to the handful of U.S. towns that have recently established pay-by-container trash-collection systems as a method to more fairly distribute the cost of what was once a cheap municipal service.

"We had a few complaints," said Diane L. Seals, the borough's deputy clerk. "But overwhelmingly we had more people saying this is a fair system and they didn't mind recycling."

Mary Keane Briggs, the former councilwoman who sponsored the measure, said the system rewards small households and senior citizens, who tend to produce less trash than younger families.

"In my opinion, most of the people who complained were accustomed to throwing out their garbage and not thinking about it," she said. "This system forces you to think about it."

Borough officials believe only a small number of people are trying to circumvent the charges.

They have heard that some residents haul their trash to work to avoid the cost of a sticker, which must be placed on garbage to prove disposal has been paid for.

A few residents have been caught disposing of their rubbish in commercial dumpsters or in neighboring towns. "A house where one person lived suddenly had 20 bags sitting in front of it," said Marilyn Hodgson, the borough clerk of Glen Gardner, which is adjacent to High Bridge.

Glen Gardner's response was to enact a similar pay-by-bag system, which goes into effect this month [January 1989]. "Their system seemed to be working well, so we're going to try it out here," Hodgson said. Several other towns near High Bridge have also imitated the system.

Word of High Bridge's success has spread such distances that officials from as far as Utah and Minnesota have called for information. The borough has received so many inquiries that it has begun charging $5 to mail information packets about the pay-by-bag system.

The borough's system was born of economic necessity. High Bridge officials were jarred into the new age of solid waste in 1987, when the Hunterdon County Transfer Station increased disposal fees from $40 a ton to $100 a ton. For that charge, the transfer station takes the rubbish to the GROWS landfill in Bucks County, Pennsylvania.

The increased fees forced the borough, which operates its trash-collection system as a separate utility, to double the annual charge to each household to $280. "When it got to be that expensive, we knew we had to do something," said Briggs.

They devised a system in which each household is charged $140 a year for trash collection, which includes 52 stickers—enough for one 30-gallon bag per week. Larger items, such as furniture, require several stickers. Residents can buy additional stickers for $1.25 each; the borough sold about 16,000 additional stickers last year.

The stickers have become legal tender of sorts in High Bridge. Residents donate them to churches and preschools. One woman was even caught with a clumsy counterfeit sticker, which she made by photocopying the borough's sticker and coloring it with a fluorescent-green felt-tip pen.

And the proper High Bridge guest always brings a sticker to a soiree. "When we go to parties, we give stickers to the hosts to help out," Bruton said.

Bruton is well acquainted with recycling. She helped recycle aluminum cans for the "I'm for High Bridge" Committee until the civic group turned over the chore to the township in September [1988].

With some effort, Bruton, the mother of seven, manages to cram the trash from her nine-member family into two bags each week. "Even little Jell-O boxes—I put them on the floor and step on them to condense them," she said.

Despite the efforts of residents like Bruton,

higher landfill costs have forced the borough to raise the disposal charges in 1989 to $200 a year and $1.65 for each additional bag. To prevent residents from disposing of trash at last year's cost, the borough changed the color of 1989's stickers to fluorescent orange.

"I heard some people say they had extra stickers at the end of the year and they were going to clean out the basement with them," Bruton said.

Indeed, it appeared that High Bridge residents did some early spring cleaning last week to take advantage of the cheaper trash-disposal costs. At some houses, the borough's garbage truck was greeted by as many as 12 trash bags.

"Some of this is Christmas wrapping paper," said Robert Cole, the driver of the truck, as he helped load a three-sticker sofa into the maw of the garbage truck. The truck took three bites to crush and swallow the couch.

Cole said he now picks up 8 tons of trash a day, compared with 11 tons a day in 1987. Still, he said, residents continue to throw away large amounts of reusable material. "A lot of people still don't know what recycling is," he said.

Angelika Bosse, who came out of her house to wheel her two empty containers into the garage, shared a similar opinion.

"This is a good thing," she said of the pay-by-bag system. "Our family came here five years ago from Germany, and I was surprised Americans put out so much trash.

"Europe had this problem much earlier, ja? And now the people here are only beginning to learn how to deal with it."

From the Philadelphia Inquirer, *January 3, 1989. Reprinted by permission.*

Earth Care Action

Corporations Learn to Turn Their Trash into Cash

By Warren Brown

Corporations nationwide are turning wastepaper into money.

AT&T last year saved $1 million in disposal costs and made a $365,000 profit by recycling its office wastepaper instead of sending it to a trash heap.

The Bank of America earns an estimated $354,000 a year, while cutting $62,000 from its annual waste-hauling expenses, by participating in a recycling program.

Similarly, MCI Communications Corp. picked up an extra $30,352 last year and saved $7,000 in disposal expenses by selling 42 tons of used computer-printout and ledger sheets to companies that turned the stuff into toilet paper and paper towels.

Corporate paper-recycling programs, once public relations gimmicks designed to placate restive environmentalists, increasingly are being used by companies that realize there is serious money to be saved and earned in the process.

Exact figures are not available for the number of companies participating in recycling programs, according to Bill Hancock, a spokesman for the American Paper Institute in New York. But he estimates that corporations recycle about 200,000 tons of paper a year, compared with almost none ten years ago.

The recycling programs are enterprises born of necessity, and an acknowledgment that the idea of the paperless office was a much-ballyhooed notion that probably will never come to pass.

If anything, the paper load is getting heavier.

Each office worker produced an average of $1/2$ to $3/4$ pound of wastepaper each working day in 1979, according to the Office Paper Recycling Service, a New York recycling-consulting group. Today, the average office worker produces at least a pound a day, the group said.

Ironically, much of that paper is being generated by computers and other electronic communications systems that were supposed eventually to make paper obsolete.

"Back in the 1970s, people were talking about the 'paperless office' and the 'checkless society,' but I wasn't buying any of that," said Burke Stinson, spokesman for AT&T's paper-recycling project.

Computers can store information for decades and zap it from one place to another in an instant, but everybody wants to hold on to a computer printout of some sort, Stinson said.

With the paper mountain growing and landfill space shrinking, disposal costs have skyrocketed. A New Jersey company that paid an average of $5 a ton for trash disposal in 1970 now pays about $100 a ton, New Jersey environmental officials say.

The disposal-price increases have created a keen interest in corporate paper-recycling programs, one that has companies from Australia to New York seeking services from the Office Paper Recycling Service.

"Ten years ago, we were a voice in the wilderness" on the issue of paper recycling, said Sheila Millendorf, director of the consulting service. "Only a few forward-thinking companies, like AT&T, were willing to listen to us."

AT&T actually started its program 25 years ago, partly as a gesture to environmentalists who criticized the company for covering the American landscape with tens of millions of its famous phone books. Now, AT&T is driven less by altruism than by profit.

AT&T's program works like this:

Workers at the telephone company's 5,000 locations across the country put dead memos and letters, along with unneeded computer printouts, into plastic desktop folders. The folders are emptied throughout the day, usually when workers go to copying machines, where huge paper-recycling bins are kept.

At the end of each business day, bins from each floor are taken to a central pickup point, where recycling companies haul them away for processing. Because some documents are sensitive, security is important.

Last year, AT&T got rid of 4,200 tons of office scrap by recycling.

In a way, said Stinson, the recycling process is a kind of poetic justice.

"We like to remind AT&T supervisors and executives that their memos and business letters eventually will wind up as toilet paper. That should be a sobering thought for many of them," Stinson said.

From the Philadelphia Inquirer, *May 2, 1989. Reprinted by permission.*

SOLID WASTE

Earth Care Action

Recycling Pioneers Are Seeking Ways to Exchange New Life for Old Tires

By Thomas J. Sheeran

Mountains of scrap tires are being eroded by entrepreneurs who are cutting them up for fuel, mixing them with asphalt to extend highway life, and using them to soften airport runways.

The pioneers are seeking profitable ways to recycle discarded tires and remove a huge trash problem. An estimated 240 million tires are added to the highly flammable heaps each year.

Some tires are recycled the old-fashioned way, by retreading, but only 20 percent are suited for that use. A few get new life as buffers for marine and truck loading docks. Some end up in the backyard as tire swings.

But most tires just piled up until business people and scientists recently stepped up efforts to find new uses.

In Cleveland, for instance, a company is experimenting with chopped-up tires as a fuel additive in coal-fired boilers to see if they can reduce acid rain, and scrap tires are being used as components in septic tank systems.

For some, finding new uses for scrap tires, "is like [hauling] garbage—you get paid to take it in and get paid to make it into a fuel," said Anderson B. Carothers, chairman of Waste Recovery Inc.

His company uses scrap tires as a supplemental fuel at three wood-processing plants.

"We compete against oil in the Northwest, against gas in Louisiana, and in the Southeast, we're competing against coal," Carothers said.

Waste Recovery spent $2 million apiece on plants in Portland, Oregon; Houston; and Atlanta, each capable of burning five million tires a year, but the company is not making money yet, Carothers said.

Robert H. Snyder, former research director of Uniroyal, a tire maker that is now part of Uniroyal Goodrich Tire Co., spent much of his career trying to build a better tire and now is working to retire the mountains of leftovers.

"It's very clear to me that a scrap tire can very well pay for its self-disposal," he contends.

Snyder, a consultant in Detroit, said much of his current research is confidential because he intends to seek patents. But he said one new use is adding chopped bits of tires to speed up the composting of sewage sludge.

"It's gathering a fair amount of momentum," Snyder said, citing community programs in New Hampshire and research at Rutgers University in New Jersey and in U.S. Department of Agriculture labs in Maryland.

If each of the 250 sludge composting sites in the nation used scrap tire chips instead of traditional wood chips to promote the process, a three-year supply of scrap tires would be needed, Snyder said.

Another trend people in the business foresee is increasing state and city regulation of scrap tire disposal. Carothers said disposal costs range from 25 to 50 cents a tire depending on licensing and transportation costs.

Florida, for instance, began a 50-cent surcharge on each new tire sale this year to finance disposal. The surcharge will double to $1 in 1990.

From the Associated Press, March 2, 1989. Reprinted by permission.

Earth Care Action

Ontario Recycling in Blue Boxes

By Fred Langan

The Canadian province of Ontario has developed one of the most advanced waste-recycling programs in the industrialized world. Individual citizens do the work, separating recyclable material from the regular trash.

It is called the Blue Box program, after the blue plastic bin in which the recycled material is placed. It was an idea generated by private businesses that knew they had to react to the garbage crisis before the government forced them to do so.

The Ontario Ministry of the Environment estimates that Ontario produces ten million tons of garbage a year, or more than one ton a person.

More than 60 percent of the people in Ontario, or two million households, can now use the program. About 80 percent of the people who have been given boxes for recycling use them regularly.

The garbage is separated at home, with the recycled material going to the blue plastic bin, including old newspapers, glass bottles and jars, metal food and drink cans, and plastic soft-drink bottles. Other garbage, including food and other household trash, is put out in cans or green garbage bags. Once a week, special trucks come by to pick up the separated garbage in the blue box. It is then transferred to a recycling center.

The goal in Ontario is to cut the amount of garbage going into dumps and landfill sites by 50 percent by the year 2000. That would be at least five million tons a year, which is equal in weight to 60 percent of the pulp and paper produced in Canada every year.

Toronto is running out of places to put garbage. Using present methods, metropolitan Toronto's two dump sites will be full by 1993. Outlying districts are not willing to be the site of the municipal dump.

There are plans to distribute backyard composters to households throughout the province. Since food and other biodegradable kitchen scraps make up 22 percent of municipal waste, widespread use of backyard composters would cut down on garbage bulk. Leaves, grass cuttings, and other yard waste make up another 15 percent of garbage.

The Blue Box program started in Ontario in 1986. It was sponsored by the soft-drink industry, steel and aluminum companies, and glass-making companies. The manufacturers were all worried that if they didn't do something, government would force them to take action. Legislation in the United States was requiring companies to do things the government way and not their way. In the 1987–88 legislative period alone, more than 2,000 solid waste bills were introduced into U.S. state legislatures. In addition, 300 measures were aimed at packaging and containers.

"The province can either take the regulatory approach, as has been done in many U.S. states, threatening to ban products that are not recyclable, or it can take the alternative approach and get everybody working together to find solutions," said Colin Issacs, an environmental activist and executive director of Pollution Probe in Toronto.

Although soft-drink containers make up less than 1 percent of the total garbage pile, they are highly visible. There was the worry that governments might simply ban certain types of packaging, such as plastic bottles or aluminum cans. Taxes on garbage might end up in general revenue and never be spent on the actual problem.

So the industry decided to tax itself to fund the recycling program. The large soft-drink firms contribute seven cents a case, or about a cent a liter on all nonrefillable containers sold in Ontario. Those include steel and aluminum cans, some glass bottles, and plastic bottles.

The soft-drink industry will provide all the funding for the Blue Box program, about C$20 million (Canadian; U.S. $16.8 million) over a four-year period. But the Ontario government wants other industries to contribute, especially the pulp and paper industry and even newspapers, which are responsible for 14.4 percent of the waste products in Ontario.

One problem is what to do with the recycled material. Waste paper is going for between C$16 to C$45 a ton, compared with the C$70 to C$100 it was worth just a year ago. Much of the waste paper is being shipped to foreign markets such as Taiwan and Japan, where there are few trees to be harvested for virgin pulp. Prices are expected to continue to drop as supply rises.

"What we have with recycled newsprint is the equivalent of an urban forest," says Harold Corrigan, chairman of Ontario Multi-Material Recycling Inc., the company that oversees the Blue Box program.

Reprinted by permission from the Christian Science Monitor, *May 17, 1989. Copyright © 1989 the Christian Science Publishing Company. All rights reserved.*

Earth Care Action

Once a Month, an Unruly City Scrapes Itself Clean

By Kenneth B. Noble

There used to be an axiom among seasoned travelers about Lagos, black Africa's most populous city: The reality is even worse than the horror stories.

It is an unruly mixture of run-down office blocks, homes made of tarpaulin and corrugated iron, and fetid open sewers. The traffic jams, or "go-slows" in local parlance, are among the world's worst, the telephone system is unreliable and telexes sputter to life only occasionally.

An official press release explaining why the government decided to build a new capital 300 miles to the northeast at Abuja said Lagos was a place of "perennial traffic jams, intolerable congestion, a chaotic sanitary situation, inadequate social amenities, an alarming crime rate."

There is a new ritual in Lagos these days, however, and it augurs the beginning of a more salutary image for the city.

Cleanup Ritual Becomes Festive

For three hours once a month, silence descends. The cacophony of thousands of car horns ceases. Shops are closed and taxis refuse to pick up passengers. The push and bustle of crowds and traffic virtually disappear. Even most local airline flights are grounded.

Then, as if summoned by an invisible drill sergeant, crowds suddenly fill the streets again, but this time it is the sounds of hammers against nails, the whisk of brooms, and the metallic clatter of garbage trucks that fill the air. For those few hours, much of the city seems gripped in a delirium of cleanliness.

The event, held on the last Saturday of each month, is known as the National Environmental Sanitation Exercise, and since it began a little more than two years ago, it has drawn ever-larger throngs of people striving to erase Lagos's reputation as one of the world's most unruly and exasperating cities.

For many, especially in a poorer section, the cleanup ritual has become a festive occasion, a reassuring sign that in spite of economic growth, there is still the ability to bring order to this often chaotic place.

Wiping Away Years of Filth

"For those three hours, virtually the whole city is out, cleaning the garden, sweeping, and scrubbing," said Frank Aigbogun, news editor of the *Vanguard,* a Lagos daily. "These days you feel odd if you're inside doing nothing."

On a recent morning along Olorunsogo Street in Mushin, a congested neighborhood of crumbling buildings, crisscrossed by open sewers, the tasks are mostly divided according to age and sex. The women sweep and tidy up, the men scrape filth from the sewers, and the children haul buckets of trash to a collection site.

Nigeria's military government ordered the event in response to a growing clamor against the poor quality of life here. And to drive the

SOLID WASTE

Residents of the Mushin section of Lagos, Nigeria, line up with refuse to be trucked away as they take part in the monthly National Environmental Sanitation Exercise. (Kenneth Noble/NYT PIX)

point home, the government prescribes fines and jail terms of up to a month for those who do not comply. So far, local newspapers say, several hundred Nigerians have been imprisoned for violating the edict.

Diplomats and businessmen—many of whom ridiculed the campaign initially as simply a propaganda exercise and doubted that three hours a month could wipe away years of filth—now agree that much of the city is markedly cleaner.

A Western diplomat who has lived in Lagos at various times in the last 15 years said, "It is sad to say, but hard times have made this a much saner, more livable place."

Aigbogun said, "This is deadly serious. The government is not joking at all about this. And it is obviously working."

But cleanliness is not the only element in the self-improvement campaign. The military government is now preparing to tackle perhaps an even thornier problem: how to stop the carnage on Nigeria's highways.

Reducing Highway Deaths

Dr. J. O. Sojobi, president of the Nigerian Road Federation, recently said the nation "should feel embarrassed" about having the worst accident record in the world. And while statistics are hard to come by, nearly everyone here can cite an example of why things have gotten out of control.

In August 1986, 72 Nigerians were killed in an automobile pileup on the Benin-Asaba Highway. The government estimates that more than 400,000 serious road accidents have occurred in the last decade.

For the moment, the government has handed over the problem to a newly created body, the Federal Road Safety Corps. The safety corps recently issued its first directive, among other measures setting the first highway speed limits in this nation's history—55 miles per hour for trucks and buses and 60 miles for all other vehicles.

Two Courses of Action

The new spirit of vigilance against dirt and disorder, like everything else in Nigeria, is associated with oil. When oil went from $2 to $34 a barrel in the 1970s, Nigeria went on a wild binge of growth and spending. From 1965 to 1980, Nigeria's national budget went from less than $300 million a year to nearly $20 billion.

Largely because the country lacked the bureaucratic sophistication to manage or even keep track of all that money, much of it was allocated to consumer imports, or a stream of wasteful products, or to political regional and ethnic patronage. All of that might have been tolerable had the oil boom continued. It did not. Oil prices crashed, and Nigeria found itself with a $21 billion foreign debt.

Nigeria's military government, which took power in a coup in August 1985, took two courses of action to rein in the years of excess: one in politics, the other in economics.

On the political front, Gen. Ibrahin Babangida, the Nigerian leader, announced plans for a phased return to civil rule by 1998. But just as important was what the army general called his determination to create "a new political generation." That became the basis for the environmental sanitation exercise.

Yesufu Manman, Gen. Babangida's press secretary, said, "We sought to inculcate the imperative at keeping tidy surroundings, and it has succeeded because Nigerians have imbibed it."

From the New York Times, *March 8, 1989. Copyright © 1989 by the New York Times Company. Reprinted by permission.*

TOXIC WASTE

Preserving Our Environment: Preventing Hazardous Waste

By Joel S. Hirschhorn

The Children's Cleanup Crusade 131
The Gift That Keeps on Giving...Toxics 136
Freiburg, the Environmental Capital of West Germany 138
Africa Gets Tough on Toxic Dumping 142

The subject of hazardous waste never seems to lose steam. Opinion polls repeatedly show that the public is more worried about toxic waste than any other environmental issue. In state elections, almost any referendum calling for action against toxic waste problems succeeds by a large margin.

Recent news about the global environmental consequences of industrial emissions, motor vehicle fumes, and other uses of fossil fuels through the global-warming greenhouse effect has focused public concern on environmental policies. While the national news media tend to ignore toxic waste issues, except for the scandals, at the local level there is a continuous stream of news stories. Contaminated drinking water, medical wastes and raw sewage washed up on public beaches, leaking landfills, transportation accidents, government regulations, and costly but seemingly ineffective attempts to clean up Superfund sites make headlines in communities across the country.

Much of the public-policy debate on toxic waste concerns siting: attempts to situate, grant permits, and operate hazardous-waste-management facilities of any kind, from landfills to

From Science for the People, *September/October 1988. Reprinted by permission. Joel Hirschhorn is a senior associate at the Office of Technology Assessment of the U.S. Congress.*

incinerators, sea-bound barges, and storage and handling facilities. The public is unwilling to accept almost any environmental risk to their health and well-being. The burden on those who want to locate hazardous-waste-management facilities is enormous. No amount of risk assessment, mediation, and economic incentives is likely to erase strong public opposition to siting facilities where they may endanger any community.

The irony is that most people fail to recognize our traditional proclivity to deal with environmental problems *after* we've created them. The history of environmental protection policy in the United States has been largely one of reactive crisis management. The environmentalism of the 1960s focused on government regulations to force industry—but not consumers—to deal with pollutants and wastes after they are created. Public policy has used one tool above all others: command and control regulations that deal with "end-of-pipe" pollution. We have been preoccupied with putting pollutants and wastes in "black boxes," somewhere between the point where they are created and the point where humans are exposed to them.

There is a sizable pollution-control industry that has arisen to help industry cope with the increasing numbers of environmental regulations. National spending on the environment has followed regulation: Today, the U.S. spends about $10 million annually for every page of federal environmental regulations. And these historic trends seem likely to continue—the more we regulate, the more we'll spend.

From Disposal to Reduction

For more than a decade, policymakers from government, industry, and environmental groups have come up with the same hierarchy for dealing with hazardous waste. In a policy statement published in the *Federal Register* in 1976, the Environmental Protection Agency (EPA) agreed that waste reduction—not generating pollutants in the first place—should be the preferred first option for waste management. The treatment, or "black-box," approach is in the middle: Using technology to convert, destroy, or detoxify waste, so as to reduce risks to health and the environment. At the bottom of the hierarchy lie land disposal and other methods of placing waste directly into the environment, including dumping it in the ocean or enclosing it in salt domes.

The irony is that most wastes have been managed in what policymakers agree to be the worst way: They've been put directly into the environment. But in the last several years, there has been an enormous public movement to limit the use of land disposal. This movement has followed, first, the discovery of uncontrolled and often abandoned toxic waste burial sites that posed threats to health and the environment and had to be cleaned up. Love Canal, the community in New York that had to be evacuated because of several health problems and deaths attributed to toxic waste, spurred the nation's multi-billion-dollar Superfund cleanup program.

Second, contaminated drinking water, discovered in the taps of more and more households and communities every year, has also fueled the opposition to waste burial. These two consequences of land disposal have created a potent public backlash toward all toxic waste facilities.

With the 1984 amendments to the Resource Conservation and Recovery Act (RCRA), Congress took explicit action to severely constrain land disposal methods, following the lead of several states. But from a public-policy perspective, we have only moved up one rung on the ladder of preferred waste-management options to the treatment or "black-box" approach. We're still avoiding the best option—waste reduction.

In the RCRA amendments, Congress directed the EPA to consider the availability of treatment capacity in its decisions to ban land disposal. The EPA was not directed to consider the ability of waste generators to reduce their waste with currently available knowledge and technology, and it hasn't taken the initiative to do so.

When it comes to waste reduction, the term used in the RCRA amendments is "waste minimization," and this term encompasses both waste reduction and waste treatment. The amendments make it clear that waste reduction is preferred. Moreover, the legislative record explicitly directed EPA not to intrude into private sector production decisions with regulations prescribing waste minimization.

End-of-Pipe Policies

There is a pivotal difference between waste reduction and both of the other two generic options—treatment and land disposal. The latter remain faithful to the end-of-pipe concept on which our entire environmental protection system was founded. Waste reduction, which everyone agrees in principle to be the best option, is fundamentally different because it prevents pollution, rather than just controlling it. Waste reduction is not simply another attempt at regulatory reform within the pollution-control framework.

Therefore, public opposition to "better" or "high-tech" waste-management options, even when they're termed "waste minimization," should come as no surprise. For example, opposition to incineration of wastes increasingly resembles the public's rejection of land disposal. Toxic pollution from incinerators still ends up in the environment.

Although treatment differs from land disposal, and in most cases is probably better environmentally, it is also similar to land disposal in several important ways.

Both treatment and land disposal require effective government regulation, but the history of the effectiveness of the regulatory system is not very satisfying to the public. Two points illustrate this: extremely high levels of noncompliance and large loopholes in which substances or practices viewed as threatening remain unregulated altogether. A good example is toxic air emissions from incinerators, which under the Clean Air Act program remain largely unregulated.

Both treatments and disposal are vulnerable to the failure of technology, people, and institutions. About 10 percent of current Superfund sites were waste-treatment or recycling facilities. Moreover, the public has become more aware of the limits and failures of even the most sophisticated technology through spectacular failures, ranging from the NASA space shuttle disaster to the chemical plant accident in Bhopal, India, and the Chernobyl nuclear reactor catastrophe. Optimism and fascination about technology may be fading, or at least becoming balanced with skepticism.

Furthermore, both treatment and disposal hold the risk of releases of hazardous substances, with uncertain effects on health and the environment. Although there is a continuing attempt by environmental policymakers to use risk assessment to educate and convince the public that specific risks may be acceptably low, there is little reliable data to plug into risk assessments.

Data inadequacies create a dependence on theoretical models that are complex enough to be incomprehensible to the public, but also so simple that they are inaccurate. The public must depend on the experts to find out what risk assessments mean, but the public increasingly distrusts experts and knows that "expert" judgments influence the results of health risk assessments.

Therefore, we come to an interesting and uncomfortable point in hazardous-waste public policy. It is based on rejecting the worst option and moving to the middle ground of waste treatment. But as we have seen, waste treatment suffers from some of the same basic flaws as land disposal.

Should we be looking for suitable backyards for waste-treatment facilities? Shouldn't we focus on waste reduction and pollution prevention in industrial plants and even in the households where the problem begins? Would investments in waste reduction yield greater environmental and economic benefits than investments in waste treatment? Is it enough to talk about controlled and acceptable risks? Or must the environmental protection system move, just as the health-care system has done, to relying more on prevention and less on reaction to environmental hazards?

Public-Policy Choices

Environmental pollution naturally resulted in attempts to use technology to keep hazardous substances from entering the environment. Traditionally, engineers had assumed that the environment could accept and tolerate wastes generated during industrial production. They made blueprints for industrial plants with arrows showing the wastes going into the air, water, or land.

After society discovered pollution in the air and water and turned to technology to control and collect hazardous or unwanted substances before they entered the environment, the land became the place to dispose of pollutants. Not until 1976 did federal public policy recognize hazardous waste to be a subset of solid waste and establish new regulations to deal with facilities that managed a broad and ill-defined range of wastes. But the RCRA program did not attempt to regulate or limit the generation of hazardous waste.

Within the primary environmental statutes—the Clean Air Act, Clean Water Act, Toxic Substances Control Act, and Resource Conservation and Recovery Act—opportunities did exist to pursue waste-reduction and pollution-prevention policies. But neither the government, industry, nor organized environmental groups took waste reduction very seriously. Instead, nearly everyone concerned with environmental protection grasped the same choice: end-of-pipe regulations dealing with pollution control.

Industry acquiescence to pollution control rather than prevention proved to be the implicit justification for this choice. Industry sought to keep the government out of its operations as much as possible. As burdensome and unnecessary as pollution-control regulations might have appeared to those in industry, they were preferable to the government dictating what could or could not be done in production or regulating the composition of products.

Environmentalists wanted immediate action: In spite of the difficulties associated with pollution control, it appeared harder to move upstream to regulate the generation of pollutants and wastes. As for the government, examination of statutes and EPA studies shows that the EPA was more interested in prevention than were other environmental policymakers. But the EPA's early interest did not find a constituency.

Only within the past few years has there been an explicit examination of prevention as a fundamentally different public-policy strategy. A series of studies, conferences, and reports have come forth on this subject from organizations including the National Academy of Sciences, Office of Technology Assessment, the research association INFORM, League of Women Voters, Environmental Defense Fund, and the EPA.

Nevertheless, although these reports are optimistic about waste reduction and detail ways to implement a pollution-prevention strategy, there has been no concentration of political activities, significant news coverage, or crystallization of the issue. There has not been a clear national movement to identify, define, and implement a new public-policy choice. Is this situation changing?

The Changing Environment for New Policies

There are several factors that could cause a major redirection of environmental public policy.

Regulatory ineffectiveness. There is increased recognition that the current regulatory system is not effective in protecting health and the environment. The result is a widespread loss of public confidence in government agencies charged with environmental protection. Increasingly, environmental problems that might have been prevented or responded to before public health was threatened now invade people's lives. These include asbestos exposures, radon in houses, toxics in drinking water from waste sites and leaking underground storage tanks, lead in products and water, and chemical plant, industrial, and transportation accidents. There seems to be a steady creation of an environmental deficit that is bound to come due. The public is also better informed today about regulatory noncompliance and loopholes, and more quickly notified about threats to their health. Like a wound that has not healed properly, the public is hypersensitive to chemical threats.

Economic inefficiency. The precarious national economy and distressed industrial base focus attention on the economic efficiency of the current system. More environmental regulations will cost society more money. The cost-benefit issue becomes more important in decision making. More people are asking questions like: Are we getting our money's worth? Are we paying too high a price for uncertain environmental protection in industrial unemployment, plant closings, and relocations to other nations?

Institutional problems. The current regulatory system is increasingly complex and difficult to administer. More and more regulations require more and more information, producing studies that still don't meet public needs. For example, more than ten years after the passage of RCRA, there is still no reliable national data base on hazardous-waste generation and management. Problems in the EPA mean that the legislative and judicial branches are increasingly doing the job of the regulatory agency. More litigation adds to delays. Actions of one part of government often negate those of another. The current regulatory system is constantly added to as existing regulations appear ineffective or incomplete—but like a soup that started out with the wrong ingredients, it never tastes quite right, no matter how many cooks try to fix it. Moreover, the ever-expanding environmental policy system grows without a strategy or enough information and resources. The federal government shifts responsibility to the states, but most states don't have the resources for implementation. Bureaucratic and institutional gridlock tightens as policymakers respond to the public's demands with more regulations.

Liabilities. The private sector faces costly and uncertain long-term environmental liabilities. These cover a wide spectrum: Superfund cleanup costs, civil litigation over injury and damage claims, criminal prosecutions of corporate officials, constraints on future sale of property because of possible toxic contamination, and uninsurable costs of accidents. Small and large companies face increasingly large, unforeseen, and debilitating liabilities, threatening the industrial strength of the national economy. Power is shifting away from national organized environmental groups that focus on federal legislation to community and grass roots activism that focuses on influencing local actions and state and local legislation. Current public policies are rejected and community fears and frustration often lead to "negative" solutions—such as blocking the siting of incinerators and other waste-management facilities. Industry asks local citizens. "What is acceptable?" And citizen activists typically answer, "That's your problem!"

Environmental activists seem disillusioned with federal policies and are more inclined to seek political solutions at the community or state level, and they have often been successful in this strategy. However, the solutions they seek are often difficult, if not impossible, to implement when they are contingent upon zero emissions or zero risk. Those solutions seem impossible to industry managers and inconsistent with the products and lifestyles of modern industrial society.

Technological pessimism. The public is growing more apprehensive about technology and the human and institutional systems that control it; technological failures get a lot of attention. Technical fixes are increasingly viewed with anxiety and suspicion; this is especially true for hazardous waste technologies. And scientists and engineers are participating in grassroots environmental activities more than ever; communities are demanding government funding to hire independent technical consultants. Indeed, the Superfund law provides such funding.

The public is asking more sophisticated questions about the nature, limitations, and implementation of technological solutions to environmental problems. These questions contribute to delays in reaching agreement on technical solutions, which are evident in the Superfund program and in siting hazardous-waste facilities. Thus, existing problems never seem to get solved, and new problems spring up. The market for new and innovative technology grows, and improved technologies are being developed to reduce hazardous-waste production in certain industrial procedures, such as electroplating and degreasing. But demands for even better technology to reach acceptable or zero risk often escalate costs and exacerbate the economic problems discussed above.

From Control to Prevention

There appears to be one fundamental shift in environmental public policy that satisfies a number of needs. That is the choice to climb the ladder of environmental protection policies to the highest and best option: pollution prevention.

No one disputes that prevention is the most certain way to reduce (and often remove) threats to health and the environment. Although it is clearly nonsense to suggest that all environmental pollutants and wastes can be avoided, there is convincing evidence that a significant fraction can be cut through waste-reduction measures that are technically and economically feasible now. Moreover, a waste-reduction effort that included all pollutants and wastes, regardless of how they are regulated now, would provide the comprehensive coverage that has eluded the current system, which was built up by addressing one environmental impact at a time.

The economics of pollution prevention are very attractive today because the current regulatory system and multiple liabilities create a stream of increasing costs that can be cut. More fundamentally, industrial spending on waste reduction increases industrial efficiency, profitability, and competitiveness. Data from the 3M company show a savings of about $300 million from 1975 to 1985 from pollution-prevention techniques that reduced waste generation by 50 percent. The savings amounted to about a 5 percent increase in 3M's net income over that period. So pollution prevention is a way for companies to cut their environmental spending even as regulations increase: 3M continues to work at reducing waste.

The institutional problems discussed above clearly would benefit from a shift to pollution prevention. Government regulatory agencies would have less to regulate. It is difficult to imagine any other action that would lighten the demands of government institutions so much, even if there are more regulations created to cover more substances and situations.

The environmental activism of today has not yet fully embraced pollution prevention as a principal strategy, in part because the media and public education have not paid much attention to this alternative. Environmentalists have been busy cleaning up and opposing immediate threat, and it has been easy to put off thinking about a new preventive strategy that can't do anything about the hazards we've already inherited.

But there is also fear that a focus on pollution prevention might weaken the current regulatory system or sap resources for its implementation and enforcement. Even though current regulations may be less effective than many of us would like, some environmentalists are apprehensive about waste reduction, especially if it is a voluntary industrial effort. And almost every study of waste reduction has pointed out how impractical it would be to use a traditional regulatory approach.

Nevertheless, there is evidence that environmental activists increasingly see pollution prevention as a way to improve protection as long as the pollution-control regulatory system remains intact. The challenge for public policy is to maintain the strength of the pollution-control system while recognizing how current regulations pose obstacles and disincentives to industrial waste reduction. For example, the current system has the force of law to channel industrial resources into end-of-pipe measures rather than to change production systems to reduce waste.

Finally, pollution prevention addresses underlying social concerns about technology. Waste reduction could help resolve siting problems by demonstrating real need for safer and fewer waste-management facilities. Waste reduction simplifies the situation by cutting down on end-of-pipe devices and by calling for a reexamination and redesign of production facilities, technologies, systems, materials, and equipment.

Some industrial experiences in waste reduction have been impressive. Du Pont claims that from 1984 to 1985 it reduced hazardous-waste production by 35 to 50 percent in some divisions. In terms of waste output relative to unit production output, Exxon Chemical Americas and Rohm and Haas both reported a 10 percent reduction from 1984 to 1985. Monsanto claimed a 20 percent reduction in waste production from 1982 to 1984, and Olin Corporation reported a 34 percent drop from 1981 to 1985.

Yet very few U.S. companies have undertaken waste reduction. An Illinois study of 275 companies found that more than 50 percent of hazardous-waste producers had not begun serious waste reduction in 1985. A New Jersey study of 22 companies found that 41 percent claimed to have implemented waste reduction between 1981 and 1985, and 36 percent said that they had plans for waste reduction in the future. A study of more than 100 small metal-plate manufacturers by a California public interest group found that 75 percent were not implementing waste-reduction procedures in 1986.

Although few companies practice serious waste reduction, neither technology nor economics appears to be the limiting factor. Instead, there is a host of human, organizational, and institutional obstacles in the private and public sectors. There has always been some waste reduction. The challenge for public policy is to find ways to increase its pace and comprehensiveness.

A Waste-Reduction Ethic

The shift in paradigm from pollution control to pollution prevention has intellectual roots in the 1970s, but it is still on the sidelines in the public-policy arena. That's because we live in a pollution-control culture: less than 1 percent of the U.S. government's environmental spending goes for pollution prevention. We need institutional champions for waste reduction and we need to change consumer tastes, practices, and products to reduce household generation of hazardous waste.

Creating a waste-reduction ethic remains a goal. For industry, that means engendering strong support, from senior management to rank-and-file employees. Waste reduction needs to become a priority of corporate culture, just as energy conservation has become a goal for many industries. Staff and management should be motivated to meet company waste-reduction goals and timetables, which can be set through waste-reduction audits. The economic motivation for preventing hazardous waste should be transferred to manufacturing and production divisions, including responsibility for future liabilities and cleanup.

Employees should be trained in waste-reduction techniques, and information should be disseminated throughout companies and industries. Technical assistance can come from state and federal agencies, private consultants, company production personnel, and environmental engineers. For example, an on-line waste-reduction data base is being developed by the EPA, and "expert system" computer programs to assist engineers in implementing waste reduction are being created by the EPA and state agencies in Illinois and Maryland.

As threats to health and the environment continue, such as more drinking water supplies found to be contaminated beyond use, an atmosphere conducive to environmental policy changes will result. Society may be driven to pollution prevention out of desperation rather than attracted to prevention because of its intrinsic benefits. If that happens, policymakers will respond with initiatives that they think have a chance of succeeding through broad public support.

Detailed policy options to implement a strong federal waste-reduction initiative have been presented for congressional consideration. In fiscal 1988, Congress allocated funds to the EPA for its own waste-reduction program and $4 million for grants to state waste-reduction programs. Legislation is also being considered that would spell out the primacy of waste reduction, define it carefully, and perhaps establish a national voluntary goal of 10 percent waste reduction per year for five years.

For the evolution of public policy, it is important to remember that the scope of current unfilled needs determines the shape of things to come. The longer we wait for public policy to embrace and tangibly support pollution prevention, however, the longer we wait for the environmental and economic benefits that it offers.

TOXIC WASTE

Earth Care Action

The Children's Cleanup Crusade

By Barbara A. Lewis

Black dots representing potentially hazardous waste sites were sprinkled across the wall map of Salt Lake City. One of the sites, the children discovered, was just three blocks from our school.

"That old barrel yard?" 11-year-old Maxine asked, shocked at how close the site was to us. "Kids climb all over those barrels."

"I bet there are at least a thousand barrels in one pile," added Chris. He grabbed a marking pen and circled the spot.

The exercise was the beginning of a spring course on groundwater that I had planned for the academically talented sixth graders at Jackson Elementary, where I teach special classes of fourth through sixth graders. I had no idea then that I was unleashing a tiger.

As it turned out, Chris had underestimated:

Jackson Elementary School students size up the dump. Some had played on the leaking drums before learning they contained residues of toxic wastes. (Paul Barker/*Deseret News*)

The site held 50,000 barrels that at one time had contained everything from molasses to hazardous chemicals. Now, after a recycling business had stockpiled them for more than 40 years, many were rusted and corroded.

As the sixth graders' enthusiasm grew, the fourth and fifth graders in my classes also became interested. I made preliminary phone calls to alert officials that the students would be calling and to ask what they might do to help. "There's nothing children can do," one state Department of Health spokesperson told me. "They'll be in high school before they see any results." Later, when the children themselves called the officials, they were shooed away like pesky flies.

Undaunted, the students conducted a door-to-door survey of their industrial neighborhood, informing residents about the dangers of hazardous waste and searching for wells from which health officials could take water samples. In a four-block area that included several abandoned houses and warehouses, the children discovered only a few wells—all cemented over—and a largely indifferent response to their information campaign. They had, however, attracted a following of television and newspaper reporters who were intrigued that youngsters had ventured into an area where adults were generally apathetic.

Before returning to school, we paused outside the barrel-site fence. Covering three blocks, the steel mountain of drums obstructed the children's view of a community sports arena, the Mormon Temple, and the Wasatch Mountains in the distance.

"Look," Maxine said, pointing. "Some of the barrels are orange and yucky."

"Rusted," Chris said.

"And some have big holes."

"Corroded," Chris corrected.

"Look at all the orange colors in the dirt," Heather said. "And black, too. I wonder if anything leaked out of the barrels."

Maxine bent down. "The fence has lots of holes in it," she said. "Bet I could climb through one." In a later survey of the school's students, 32 children would admit to having played on the barrels.

"I've seen bums build fires in those barrels," one child said now.

"Chemicals in the bottoms could cause an explosion," another added.

Kory's brown eyes popped. "And blow up the whole school."

"Don't exaggerate," Chris scoffed.

While we were at the site, we decided to stop at the barrel yard's office. A worker denied that there were any problems at the site. A newspaper reporter who had remained with us questioned the man and was told that no chemicals from the drums had leaked onto the soil; furthermore, the owner had never accepted barrels that had more than an inch of residue in the bottom. The owner had spent more than $50,000 in the past few years to meet regulations set by the Environmental Protection Agency (EPA), the worker insisted.

The children were not convinced. "What happened to the ground before we had those laws?" Heather asked later. To learn more, they began reading articles on hazardous waste in magazines, and I obtained more information from a newspaper-clipping service. The children then wrote papers on related topics. I invited an environmental consultant, health officials, and Salt Lake City's emergency hazardous-waste cleanup team to lecture in class. One health official told us that even a single inch of chemical residue in the barrels could leak into groundwater and contaminate it.

With newly developed expertise, the children brainstormed solutions to the problem. Their first major success was gaining the support of Mayor Palmer DePaulis during a visit to his office. Their recommendation: Remove the drums and test the soil and water for contamination. DePaulis, a former teacher, promised to work toward cleaning up the site within 18 months.

In a few weeks, changes began. Utah Power and Light, which owned the land, refused to allow the owner of the barrel yard to renew his lease on the land; he retired and sold the business to a California-based barrel-recycling operation. DePaulis pressured the new owner to move the barrels, while media coverage increased awareness of social responsibility within the community.

Removal of the barrels began immediately. At a cost of a dollar a barrel, the new owner would remove more than 37,000 reusable drums during the next year, sending most of them to its California plant. The former owner would pay to remove the disintegrating remains of nonusable barrels and to send toxic materials to a hazardous-waste disposal facility.

In early June 1987, just a few months after the children had begun their campaign, researchers from the Denver EPA office came to Utah to collect soil and water samples at the barrel site, taking operation of the investigation away from local health officials. The EPA officials promised test results by fall.

Many of the children were attending their sixth-grade graduation party the day the EPA started work, but they chose to leave it so they could watch as the water and soil samples were collected. "Kids can make a difference," they began saying—kids from Jackson, which has the lowest income per capita and one of the largest minority populations in the Salt Lake City School District.

Although the project slowed for summer recess, the children called me at home to tell me what was going on at the site. In the meantime, I received two phone calls threatening that parents would take legal action against the school if the children remained involved. I suspected the calls were a ruse to scare me—parents had submitted permission slips for every excursion—and I questioned each parent to confirm my theory. All were highly supportive and eager to have their children participate. Our principal, Pete Gallegos, encouraged us to forge ahead, and the school district promised legal help if we needed it.

When the children returned to school in the fall, I expected their enthusiasm to have waned, but they immediately proved me wrong. I invited last year's sixth graders back to brainstorm additional strategies with the new sixth graders, and when I drove to the junior high to pick them up, 10 of the original 14 piled into my compact car.

I covered the chalkboard with their ideas: Write to the Denver EPA to check for test results; contact local health officials; call the mayor, the power company, the new owner of the barrel yard. Removal of the barrels had slowed down, so the children suggested applying more pressure. Do more research, they said.

I asked the children to think about who else has rights that need to be protected.

"The barrel yard owner," Chris said. "What's going to happen to small-business owners like him, who can't afford to clean up their messes? They could lose their businesses, and then only the big guys would still be around."

"But we have a right to know what's in that dirt," Heather insisted. "We're living by it." The other children agreed.

"Then who should be responsible for cleaning up hazardous waste?" I asked.

"Maybe the health department," a child said.

The students pose on the steps inside the Utah State Capitol with State Representative Ted Lewis. Lewis was one of the sponsors of legislation proposed by the students and passed by the legislature to help clean up hazardous-waste sites in Utah. (Gary McKellar/Deseret News)

But a health department spokesperson had already told us that the agency had no money for cleaning up the waste.

"Let's earn some money to help everyone, like small businesses and people like us," one child suggested.

"And give it to the health department," another added.

The children decided to hold a white-elephant sale, and they raised $468.22—not a hefty sum, but enough to clean up one square foot of a mound of toxic mess. Heather remained optimistic. "Big things can happen in small steps," she said. "Like climbing mountains."

By Christmas the long-awaited EPA test results were announced publicly. Heather tore down the hall from the office, waving a large manila envelope.

We flipped through the pages together. The report indicated that harmful chemicals, solvents, coal tars, pesticides, and heavy metals had contaminated the soil and groundwater. It listed such substances as benzine, toluene, lead, zinc, and copper. A health official visited our class to help us understand the results.

The EPA, which was now investigating the surrounding neighborhood to determine how far the contamination had spread, promised additional test results within a year. After that, a Utah health official told me, the site would probably go onto the EPA's Utah Priorities List of areas slated for cleanup.

By last April [1988], all 50,000 barrels had been removed, and the EPA began pressuring the yard's new owner to build a fence that would effectively keep out transients and children. Already my students had warned the entire school to stay away.

Meanwhile, the children had received and accepted invitations to speak before a number of organizations. Their next step was to mail out 550 letters to industries, environmental groups, and service organizations to seek additional funds for cleanup. We received stacks of congratulatory letters and checks. Only two negative letters (from industries) arrived, and we got one discouraging phone call. No further threats were made.

Altogether the children raised more than $2,200 in cash and another $500 in promised services. They were disappointed to learn, however, that they could not legally contribute the money to the state and have its use monitored. They didn't want their money wasted in court battles; they wanted it to go directly toward cleanup of some site.

Having read about the national Superfund, which helps clean up abandoned toxic-waste sites, one child suggested a law that would create a similar state fund, to which anyone could contribute. Utah had no such fund and ranked 45th among the states for developing environmental programs, according to "The State of the States," a 1987 study published by the Fund for Renewable Energy and the Environment.

Each fifth and sixth grader involved with the project wrote legislation, and the children combined the best points. The result was a draft bill to set up a state contributory Superfund that would not require state tax revenues. Utah House Reps. Olene Walker (R) and Ted Lewis (D) cosponsored the bill.

April, a fifth grader, then contacted Rex Black (D), who agreed to sponsor the bill in the state Senate. The children lobbied in person, handing out fliers with red-crayon trim. When they looked out the window of the capitol building, they could see in the distance the barrel site where their odyssey had begun almost a year ago.

The children were clearly enjoying themselves, but they were also aware that this was serious business: They had to reach each member of the Utah House and Senate. Back at Jackson Elementary, they tied up the phones for two days as they called each legislator. Later five children addressed the Utah Senate and five others spoke before a House committee. They received a standing ovation.

As the final votes were tallied, the children sat quietly counting in the Senate gallery. House Bill 199, a state contributory Superfund, passed without a single dissenting vote. Not allowed to applaud in this formal setting, the children grinned, mouths open in silent cheers, arms waving wildly.

"These children did something we couldn't do because 'Superfund' is such a political issue," Brent Bradford, director of the Bureau of Solid and Hazardous Waste, told the Women's State Legislative Council. "No one thought to tell the children they couldn't succeed. They got people who wouldn't even speak to us to talk to them and even to donate...they've raised the level of awareness of the whole valley to hazardous-waste issues."

Last fall, a year and a half after they began their project, the children applied for a Waste Minimization and Recycling Grant, attaching it to a Department of Health proposal, which could increase their $2,200 to $22,000 with a 10-to-1 match from the EPA. The health department's proposal was for a match to implement a program for reducing waste at its source and for encouraging recycling. In addition, legislators are considering appropriating sizable funds in their next legislative session for cleanup of toxic waste.

What next? That depends on the children. Already they have proven that kids can solve problems, as Heather says, "one small step at a time." If they can climb mountains, one would hope adults can, too.

From Sierra *magazine, March/April 1989. Reprinted by permission.*

Earth Care Action

The Gift That Keeps on Giving... Toxics

By Peyton Curlee

WARNING: The batteries that brought the flash, whir, and jingle to your holiday toys contain heavy metals and chemicals that can be harmful to the environment and human health. The heavy metals found in household batteries—mercury, lead, lithium, manganese dioxide, zinc, cadmium, silver, and nickel—can leach into soil and groundwater when buried in landfills or end up as toxic air pollutants and ash when incinerated.

The health threats of each of the primary metals in batteries are well documented: Occupational Safety and Health Administration (OSHA) workplace exposure limits are set for all eight, drinking water levels are regulated for four under the Safe Drinking Water Act, and open-air levels are restricted for two under the Clean Air Act. The big ecological risks of tossing these tiny batteries become clear when the small amounts of toxic metals are multiplied by the 25 million household batteries that are used each year in the U.S. alone. In Japan, more than 23 million "button" batteries for cameras and watches (each with several milligrams to 1.1 grams of mercury) are manufactured annually, reports the Spring 1988 *Japan Environment Review.*

Batteries are responsible for a high percentage of airborne mercury pollution, according to the Swedish Environmental Protection Board, which found that 35 percent of background levels of mercury in Sweden were attributable to the incineration of batteries. A battery burn-ban and upgraded pollution-control measures at one Stockholm incinerator reduced mercury emissions by 80 percent. The Quebec-based environmental consulting firm Eutrotech Inc. has determined that batteries are also the principal source of mercury emissions in Canada. The New Hampshire/Vermont Solid Waste Project has told the recycling industry monthly, *Resource Recycling,* that "Excluding all types of batteries from municipal solid waste would have a major impact on the amount of mercury, cadmium, and to a lesser extent, lead, in the emissions and ash residues of municipal waste-to-energy plants. An added benefit...would be less zinc in the effluents."

In 1983, Tokyo's Research Institute for Environmental Protection reported that the burning of batteries at one incinerator on the outskirts of the city was responsible for mercury emissions that were 5 to 30 times greater than levels recognized as safe by the World Health Organization. Public pressure eventually forced local governments to separate batteries from garbage for incineration, though the recovery rate is only around 5 to 15 percent, according to the *Japan Environment Review*.

The Environmental Protection Agency (EPA) has done little in the way of research or regulation of battery disposal. In September 1988, *Environmental Action* magazine charged "at least one agency staff person is accusing EPA of criminally sitting down on the job." Dr. Cate Jenkins, an EPA chemist, has called for a congressional oversight investigation of numerous alleged "unethical and criminal acts" by EPA officials in the classification of hazardous wastes, including lithium batteries.

Prospects for Recycling

Recycling the hazardous metals in batteries is one way to keep them out of the solid waste stream and environment, but currently markets exist only for those batteries that are sold in the smallest volume—silver oxide and mercury oxide button cells and nickel/cadmium rechargeable batteries. One process used to reclaim the mercury and silver from button batteries involves thermal treatment, condensation of the resultant gases to retrieve the mercury, treatment of the intermediate product with nitric acid, and finally dilution with hydrochloric acid to precipitate out the silver and zinc.

Recycling of alkaline and nonalkaline standard cells should be explored aggressively, since by 1990 they will make up about three-quarters of household battery sales. Voest-Alpine Environmental Engineering of Austria has developed an experimental process that, with pressure, magnetic screening, and thermal treatment, removes the mercury and zinc and results in an ore substitute that can be used in the manufacture of electrolytic manganese dioxide.

The most extensive attempt to undertake battery recycling—a pilot project in Japan—has not had encouraging results. Despite the participation of 73 percent of towns and cities and 100 percent of retail stores, recovery rates are low and the program costs $525 per ton, plus $40 to $230 per ton for transportation. This has not convinced U.S. battery manufacturers that recycling is feasible or desirable.

Consumer pressure and the threat of costly recycling has forced advances in lowering the quantity of mercury in batteries as much as 71 percent since 1981. Consumer pressure has also created a booming market for the new nickel/cadmium and lithium rechargeable batteries that can be recharged up to 1,000 times and have a shelf life of ten years. Lithium, however, is a soft, highly reactive metal that can burst into flames and produce explosive vapors in ordinary situations. Manufacturing requirements currently limit lithium batteries to the

9-volt size, which makes up a shrinking 8 to 10 percent of the U.S. battery market.

Community Programs

In October 1987, the people of the Danish island of Bornholm became the world's first community to put a refundable surcharge on batteries. A fee of 70 cents was placed on all batteries sold, refundable upon return of the used battery.

In 1985, the Japanese government, working with the Battery and Electrical Appliance Industry Association, installed collection boxes in 111,100 appliance and camera stores in Japan. Unfortunately, only button batteries are collected in these boxes.

Meanwhile, in the U.S., the New Hampshire/Vermont Solid Waste Project has managed to collect 1,000 pounds of batteries in the first year of its 70-store collection project. Store managers assist by encouraging customer participation and awareness of the solid waste problem. Volunteers separate mercury and silver oxide batteries for recycling and send the rest to hazardous-waste landfills.

Similar drop-off programs have been instituted by Environmental Action Coalition of New York City, the Household Hazardous Waste Project in Springfield, Missouri, and the American Watchmakers Institute. In San Francisco, the city's recycling project began accepting batteries at collection centers early in 1989.

What You Can Do

Here are some steps everyone can take to help reduce the potential problems from all kinds of batteries.

- Get the word out about the hazards of battery dumping.

- Use rechargeable batteries.

- Encourage industry to further reduce levels of toxic metals in batteries.

- Push for laws requiring safe disposal of batteries, with a refund incentive.

From Earth Island Journal, *Winter 1988/89. Reprinted by permission.*

Earth Care Action

Freiburg, the Environmental Capital of West Germany

By Petra Losch

Freiburg, West Germany, a city in the midst of a beautiful landscape of mountains and little hidden valleys, is rich in romantic and traditional buildings whose historic walls contain the spirit of almost eight centuries. But this picturesque city, which is justly famous as a major tourist attraction, is also known for its alternative power projects, its liberal spirit, and its innovative initiatives to protect the environment.

Located in the southwestern corner of West Germany, a stone's throw from France and Switzerland, Freiburg has acquired a reputation for its environmental politics and progressive municipal programs. As one of the cities boasting the greatest number of urban bikeways, an extensive recycling system for garbage, a municipal Department for Environmental Affairs, and an average vote for the Green Party of 18 percent, Freiburg seems to be the land of milk and honey for social innovators and urban environmentalists.

In 1985, Freiburg became the first city in the southern state of Baden-Württemberg to establish a Department for Environmental Affairs. A special feature of this department is that it is not a rigid, out-of-reach administrative apparatus; on the contrary, it has its own telephone "hotline" (216-3890) that gives every citizen the opportunity to call the department to ask questions or air complaints.

The "Eco-phone" was created in 1986 and has served as a connecting link between the public and the local government. Ten to 15 calls a day reach the department over this phone link, though this number rises enormously after events like the Chernobyl nuclear reactor explosion or during smog-causing temperature inversions. In the wake of Chernobyl, citizens called to ask what levels of radioactivity had been monitored in their food and for instructions on precautions they could take to safeguard themselves from exposure.

Number one on the "top ten" list of questions concerns disposal problems for pollutants and special household refuse—i.e., "Where can I dispose of waste oil after changing the oil in my car?" or what to do with toxic hazards in the home such as the formaldehyde vapors that develop after painting. Freiburg officials provide free formaldehyde-level checks (a service that other cities charge $200 or more to undertake).

The runner-up on the charts includes the filing of complaints against "environmental criminals" who have been caught red-handed by vigilant citizens. The Eco-phone was in the headlines recently when citizens discovered a red, foamy broth in a local aqueduct instead of the usual clear water. The delinquent was unmasked and turned out to be a nearby heating power plant.

It goes without saying that the Eco-phone is quite a success. "The calls indicate where the really 'hot issues' of our work are," explains Dieter Worner, the head of Freiburg's environmental department. "It also reveals what our citizens know about nature and the environment."

Another step in the direction toward a healthier and cleaner environment is the creation of an extensive municipal transit system. The most famous example (which was copied by a lot of other city administrations) was the institution of the *Umweltkarte*—the "environment ticket." The green *Umweltkarte* is of special interest to commuters since it can be used interchangeably on all city bus and trolley systems. It is also cheaper than all other monthly transit tickets. Used in conjunction with "park-and-ride" system, *Umweltkarten* have helped reduce the enormous environmental stress the city has to face, with thousands of commuters streaming into the city's limited parking space. In addition to the fact that almost every suburb can be reached easily by bus or tramway systems, Freiburg has built one of Germany's most elaborate bikepath systems, reflecting the needs of the 30,000 students whose favorite form of transportation is still the bike.

Although Breisgau County (of which Freiburg is the capital) has enormous problems with waste disposal—the two waste-disposal sites will be filled to capacity in 1989 and 1995, respectively—the city itself and the surrounding villages provide an exemplary recycling system.

Every household has two separate trash cans—a gray one and a green one. (Apartment houses have large common bins.) Green containers are for recycling garbage, including paper, glass, aluminum, clothes, wool, cardboard, and even plastic. Nonrecyclable and organic wastes (which in some cities is sorted into brown containers) belong in the gray cans.

Refuse is collected twice a week and taken to a refuse utilization plant, where the contents of the green containers are sorted. Many of the communities in the area have not yet distributed green containers, where everybody can deposit recyclable material, even special refuse like waste oil or used batteries.

An extensive removal system is almost traditional in Freiburg. Since the Middle Ages, the core of the city has been laced with a network of little aqueducts, which have contributed a special atmosphere to the city's life. Used in past centuries as a drainage system, the former "waste-quaducts" are now restricted to use as a tourist attraction. Ancient anecdotes are still attributed to these aqueducts, one of which says if you step into one, you'll marry a Freiburg resident.

A place for liberal politics, alternative culture, and a strong Green Party, Freiburg is also a breeding ground for many nonprofit organizations, the most important of these being the BUND (*Bund fur Umweltund Naturschutz Deutschland*—i.e., the German Association for Environmental and Nature Protection). The activities of the BUND comprise teaching programs, in cooperation with the local schools; field trips; monitoring endangered species; and working to establish biotopes. The BUND members also promote projects such as a recently published investigation of air chemistry and radioactivity—a problem that is of great importance for Freiburg because of the nearby French nuclear power reactor in Fessenheim, a plant notorious for accidents and poor safety standards.

The BUND was very successful in canceling a traditional motorcycle race that had been held at the *Schauinsland,* one of the Black Forest's most famous and most beautiful mountains. From the *Schauinsland* (which literally translates as "look into the country"), it is possible to view the Swiss Alps, three hours away by car. The *Schauinsland,* like large portions of the Black Forest, is threatened with extinction.

The BUND attracted attention all over Europe in 1988 when it organized a large rally of American and Canadian Indians and Australian Aborigines to protest uranium mining, which threatens the land and lives of the native peoples. The BUND conference revealed that a high percentage of groundwater has been poisoned by accidents in the mining plants around the world.

Another Freiburg institution is the *Oeko-Institut* (the Eco-Institute), an independent organization consisting of respected scientific experts in the fields of chemistry, biology, ecology, physics, agriculture, forestry, engineering, and other disciplines. The institute is known nationwide for its excellent, critical, and carefully thought-out studies and reports. In 1987 the institute started an investigation on behalf of the federal government about the risks and benefits of nuclear power compared with alternative energy options. This produced one of the institute's most famous studies, a report on the "energy inversion," which proved that by using conventional and alternative energy, European countries would be able to renounce reliance on nuclear power. Any person, company, or organization can be a member helping to support the *Oeko-Institut.* The most prominent and important member is the city of Freiburg itself.

Among the many environmental initiatives

in recent years, two major events are worth mentioning: "Ecomedia '87" (a week of international ecological films) and "Eco '88" (the largest environmental exhibition in Europe, presenting alternative technologies ranging from solar and wind power to organic food and the latest recycling equipment).

A lot of attention was paid to the work of the first federal congress regarding the creation of the city-level environmental impact reports—the *Umweltvertraglichkeitsprufung,* or UVP for short. The congress, held under the auspices of the Federal Minister of Environmental Affairs Klaus Topfer, drew some 400 experts on politics, government, and commerce to discuss questions of communal environmental protection. The result is that every planned structure in a city—buildings, streets, new highways—must now be investigated to determine its effect on the environment. Guidelines by the European Economic Community dating from 1985 state that the law must be in effect by 1988. A lot of communities have already established a UVP policy on a voluntary basis.

In the last two years, Freiburg has banned the use of pesticides in all its urban areas. The city has also instituted an "early smog warning system," which warns high-risk groups of hazardous levels of SO_2 and NO_2.

What are the reasons for all the various activities that make Freiburg so exemplary compared to other cities? Certainly the liberal politics of the city, the open spirit, and the alternative movement fed by the energies of the 30,000 students who comprise a major segment of the 200,000-member population are factors. They provide fertile soil for these initiatives. But there is also the experience of living in a beautiful, historical place that makes people more conscious of the value of a healthy environment and makes them wish to conserve what they have. (I think this is the same reason that California is one of the U.S. states most concerned with protecting its own beautiful landscape.)

On the other hand, the city and country have to face a lot of problems, and perhaps this leads to the present status: the enormous amount of waste with no possibility of disposal, dying forests, accidents in the Sandoz chemical plant and the Ciba-Geigy plant in Basel, Switzerland, that turned the Rhine river into a poisonous broth, as well as nuclear power plants in France where security seems to be of minor importance.

The resistance to nuclear energy is not restricted to that portion of the city's population that is part of the academic community. A small village near Freiburg, named Whyl, has become famous for its powerful resistance to a planned nuclear power plant and for having forced the state government to drop the plan. This is one decision for the environment which should encourage further endeavors and developments in this city. There are still a lot of unsolved problems, a lot of things to do for a healthy environment, but, in comparison with other cities, Freiburg shows the way to go and remains an inspiration for others to follow.

From Earth Island Journal, *Winter 1988/89. Reprinted by permission.*

Earth Care Action

Africa Gets Tough on Toxic Dumping

By Kingsley Moghalu

Along with the international push to curtail toxic-waste dumping, Nigeria—a victim of such dumping last summer—is taking steps to prevent a recurrence of such incidents. The government has already enacted a Harmful Waste Decree, which stipulates life imprisonment for anyone found guilty of dumping or aiding the dumping of toxic waste in Nigeria.

The punitive measures are part of an effort by African countries to tackle the rising incidence of toxic garbage "imported" by the industrially advanced countries. All 16 members of the Economic Community of West African States now have such laws. But, at an international meeting this month [March 1989] in Basel, Switzerland, Nigeria was among 40 member states of the Organization of African Unity that refused to sign an agreement on toxic-waste dumping. These states want an outright ban on waste exports to Africa.

Until last summer, most Nigerians were unaware of the existence and effects of chemical wastes. But news media reports revealed that between September 1987 and May 1988, an Italian firm, with the help of corrupt Nigerian port officials, illegally shipped 10,000 barrels of radioactive nuclear waste products. These products were deposited in an empty lot at the small Nigerian port of Koko.

Italy has evacuated the barrels, but Nigeria has brought legal action against it in the World Court, claiming damages on the basis of state responsibility in international law.

The Koko case is only one of several that have occurred in West Africa in recent years. Strapped for cash and burdened by debt, some African countries have accepted waste drums for handsome fees. But since the uproar, most have abandoned such agreements.

Early in 1988, Guinea-Bissau canceled a contract to accept 12 million tons of toxic wastes over five years for $120 million a year, a sum that favorably competes with that nation's annual gross national product of $150 million.

In February of the same year, a Norwegian shipping company—Torvald Kalveness—dumped 15,000 tons of "raw material for bricks" on the island of Kassa in Guinea. Suddenly, trees and plants in the island started dying. Investigations showed that what was said to be raw materials for bricks was actually toxic incinerator ash from Philadelphia. Authorities arrested Norway's consul general, who authorized the importation of the waste.

According to Kaare Borch, an official of the firm, "Their reaction is excessive. They've been fooled by ecologists whose goals are not realistic in our industrialized world."

But the Africans see it otherwise. In the words of a May 1988 resolution adopted by the Organization of African Unity: "We declare that the dumping of industrial and nuclear wastes in Africa is a crime against Africans, and we condemn all companies that participate ... in introducing these wastes into Africa. We ask them to clean up the areas already polluted."

From the Christian Science Monitor, *March 30, 1989. Reprinted by permission of the author.*

TROPICAL FORESTS

Fire at the Equator

By Russell Wild

Lovely Cheetahs, Meter-Saved 150
For Africa: A Fast-Growing Forest 151

Francisco Mendes Filho's reward for saving Brazilian rainforests was an assassin's bullet. His death on December 19 [1988] violently symbolizes the battle over rainforest destruction in the tropics. Mendes, a Brazilian ecologist and union organizer of that country's rubber tappers, was gunned down outside his home by a killer allegedly hired by the son of a prominent cattle rancher. Mendes was one of the many casualties in an escalating war involving not only ecologists, rubber tappers, and cattle ranchers, but also multinational corporations, huge banks, and world governments. The fate of the rainforests, which many scientists believe play a role on earth as important as spark plugs to an engine, is at stake.

Commonly called jungles (a term that tends to make scientists cringe), these tropical forests are arguably the world's grocery store, seed catalog, medicine cabinet, and air conditioner—as well as the habitat of more than half (some believe *considerably* more than half) of all life species.

"If the destruction of the rainforests continues, the earth will suffer the most devastating blow to life in all our history," says Thomas E. Lovejoy, the assistant secretary for external affairs at the Smithsonian Institution.

From Organic Gardening, *May 1989. Reprinted by permission.*

The rainforests presently cover 6 percent of the planet's surface. About half of their original turf has been lost, mostly in the past 40 years. What remains is an area roughly three-quarters the size of the continental United States. They are being destroyed so fast—at estimates of a mind-boggling 100 acres a minute, or almost a football field a second—that if present rates of destruction continue (or accelerate), there'll be mere patches left within 20 years, says preeminent British naturalist Norman Myers, Ph.D. Most important, that which we lose will be largely irreplaceable.

Brazil's annual deforestation of 20 million acres (an area the size of Indiana) makes it the world's leader, but not all of the deforestation is occurring in Brazil. Nor is the destruction all the result of cattle ranchers or other wealthy businessmen exploiting nature for fast profits. The issue is far more complex than that. In many cases, the culprits are often forced into that role. They are more like Eulalio Garcia Garcia and his family.

Amidst the sounds of large, black monkeys that bark like dogs (the locals call them *congos*) and the chattering of exotically colored, large-beaked birds in the towering trees beyond his home, Garcia toils over his small plot of land located on one of the last virgin rainforests in Costa Rica, unaware that he is the focus of a raging international controversy.

Garcia feeds his wife and seven children with the bananas, avocados, cassava, and rice that he grows on his land and with the measly profits received for the timber he cuts and sells. They live in a small house with dirt floors that he built with some of this same timber. Garcia and his family are *parceleros,* or squatters, which means if they can successfully live off the land, within six months they will gain certain rights to ownership under Costa Rican law.

"Life is hard here, but I am happy because my children will have land," he says. Garcia complains that the soil is very poor and acidic and that he cannot afford fertilizer. But despite the hardships, he and his wife plan to have more children. How many more, they're not sure. Meanwhile, their oldest son will soon be ready to start a family—and to make another clearing in the increasingly diminished rainforest.

While it's evident something must be done to save the rainforest, even the most ardent environmentalists agree that the Garcia children must have enough to eat. "Conservation cannot occur without taking account of man. Man is part of the ecosystem," says Rodrigo Gamez, the adviser to Costa Rica's president on matters of natural resources.

From his humble office, Gamez looks out over the tops of palm trees and terra-cotta roofs to the green mountains that lie several miles beyond this capital city of San José. Beyond the mountains lies Costa Rica's most treasured natural resource: the rainforest.

"There is gold in one area of Costa Rican rainforest," says Gamez. There's probably enough gold to pay off Costa Rica's crippling external debt. But when prospectors invaded the park in 1985, threatening to upset its trees and animals, the already-bankrupt Costa Rican government spent $3 million to entice the intruders out. "In this country," says Gamez, "nature is considered more valuable than gold."

The Central American country, two-thirds the size of West Virginia, with a per capita annual income of less than $1,400, is struggling to protect its endangered rainforests. Gamez and others like him question whether it's fair that little Costa Rica should shoulder the entire burden, especially since these natural treasures benefit the entire world.

"If the rapidly vanishing rainforests are to survive, we in the temperate climates must re-

Using basic technology, natives of Peru's Palcazo Valley are learning to use the rainforest without destroying it. (William Cordero)

alize that their destruction is very much our problem, too," says Peter Raven. Ph.D., director of the Missouri Botanical Garden and one of America's top experts on tropical rainforests.

Tropical rainforests are found in South and Central America, Australia, southeast Asia, and central Africa, generally within a 3,000-mile-wide belt straddling the equator. Graced with limitless sunshine and rainfall, they have evolved over millions of years into warehouses of nature capped by the plush green canopies of 100-foot trees.

To begin to understand what such a loss would mean, consider that just one square mile of rainforest in Peru hosts 1,450 species of butterflies, while the entire United States and Canada together host just 730. Consider that more species of fish live in the Amazon Basin's Rio Negro than in all U.S. rivers combined. And consider that Panama, smaller than South Carolina, has as many plant species as all of Europe.

A typical four-square-mile patch of rainforest teems with a vast range of life forms, including 125 species of mammals, 400 species of birds, 100 species of reptiles, and 60 species of amphibians, as well as 1,500 species of plants, according to the National Academy of Sciences.

Farmers are largely responsible for the fires that burn along the equator, blackening huge expanses of forest and sky from Brazil to Indonesia. What destruction is not attributable to "slash-and-burn" agriculture is due mostly to lumber cutting and cattle raising.

In part, the factors that lead farmers, loggers, and ranchers in the tropics to topple forests are the same ones that led farmers, loggers, and ranchers in 18th- and 19th-century America to do the same: survival and a better life. But there are huge differences between the two worlds. Today's Brazilian farmer who torches the Amazon Basin might also be compared to the modern-day American farmer who uses excessive amounts of environmentally harmful chemicals.

His actions, says Raven, are partly based in economics, ecologically unsound and short-sighted government policies, businesses and banks operating only for short-term profit, and in many cases, a perceived lack of alternatives.

A Global Problem

It's clear to both Raven and Gamez that the rainforest will be saved only with assistance from the wealthier, more technologically advanced countries in the Northern Hemisphere. But what must come first, they argue, is public awareness of the rainforest's extreme importance to everyone *everywhere*.

One of the most immediate dangers of tropical deforestation is the role it is playing in the warming of the planet, an increase in the so-called greenhouse effect, already beginning to wreak havoc on coastal cities and world crops.

Every year some 5.2 billion tons of heat-trapping carbon are released into the air from burning fossil fuels, and another 1.8 billion from burning tropical forests. If, however, the present rate of deforestation continues, the carbon released from the forests alone soon could reach 5 billion tons, says Myers. (At present, between 2 and 5 percent of the earth's surface is set aflame each year, chiefly due to slash-and-burn agriculture.) Moreover, the trees' ingestion of carbon dioxide and emission of oxygen is crucial to our environment.

Recent studies now show that months after these fires subside, the burned soil continues to emit high levels of nitric oxide and nitrous oxide, gases that destroy ozone and contribute to acid rain. But the climate linkages don't end there, says Myers. The rainforests contain much of the planet's moisture and absorb much of the sun's rays. "When the rainforests disappear," Myers says, "we will find that climate changes swiftly ensue, some of them so abrupt and sizable that we will have little time to adapt to their arrival."

Agriculture has much to lose from the destruction of the rainforest, yet farmers are leading the destruction. Adding to this irony is the sad truth that most rainforest soil is pitifully poor, and farmers trying to raise crops in the rainforest typically fail miserably.

It's deceiving, though. So lush is rainforest vegetation that one assumes the soil beneath is luxuriant. But most rainforest soil is in fact exhausted, old, and long devoid of nutrients. The natural vegetation thrives only because it has adapted to feeding off the decomposing organic matter constantly littering the forest floor.

Burning the forest is not only a quick and easy way of clearing the land, but the ashes supply enough nutrients for one, two, maybe three years' crops. Then the land is typically abandoned, and the farmer must burn more virgin forest to feed his family.

The law of supply and demand is not on the natives' side. Rapidly growing populations and massive unemployment in tropical countries with minimal resources leave little land—even the most forsaken—from which to eke out a living. Government policies in many of these countries only make matters worse. Some governments *subsidize* deforestation. They do this to please special interest groups such as cattle ranchers, to finance high-visibility projects such as dams for hydroelectricity, and to earn hard currency with which to pay off their crushing international debts.

Encouraging settlers to colonize the forest (even though the government knows they are doomed to fail) also provides an escape valve from the pressures of the swarming, jobless cities.

Such government short-sightedness and neglect of natural resources, however, eventually backfires. Haiti has more thoroughly deforested its land than any other nation in the Western Hemisphere. With no more forest to clear, the nation's agricultural sector is bust and its need for timber is critical.

Costa Rica's Answer

Not all governments of tropical countries neglect nature for short-term economic gain. Costa Rica should be the model for rainforest policy reform.

"Costa Rica is certainly one of the countries that has done the most for rainforest conservation," says Alvaro Ugalde, superintendent of Costa Rica's first national park and director of the country's park service from 1974 through 1986. Under his guidance, almost 11 percent of the country has been designated as parkland, and another 13 to 14 percent has been placed under some kind of conservation control.

With such dedication, one would think that deforestation in Costa Rica is a thing of the past. But in 1950, roughly 80 percent of the country was forested; today, 80 percent is deforested. The rate of destruction has not abated; nothing has changed in nearly 40 years. "The disappearing forests have brought a lot of tears to my eyes," says Ugalde, who predicts that within 8 years, all of the forests outside the parklands (and perhaps even they) will be gone. "I don't know if anyone has it in his power now to stop deforestation," Ugalde laments.

Costa Rica's conservation efforts are severely hampered by a lack of funds. With an external debt of $4 billion (the third largest, per capita, in the world), the government can spend only $1.5 million a year on the maintenance of its parklands.

Charles Schnell, resident director of the Organization for Tropical Studies in Costa Rica, says that money is indeed a problem. The government lacks sufficient funds to effectively enforce logging restrictions, and lumber companies can bribe government inspectors. "Lumber-truck drivers can pay more in bribes on a single trip than the inspectors earn in a month," he says.

The cutting of lumber (which in Costa Rica is recognized as the leading threat to rainforests) is government regulated. Yet caravans of flatbed trucks loaded with chained-down logs are seen chugging out of rainforest areas at odd hours of the night. The Costa Rican people speak of *chorizos,* or "scams." The Costa Rican Ministry of Natural Resources prohibits nighttime trucking.

The growing number of homeless also is a problem. "Colonization is synonymous with deforestation," says Gary Hartshorn, Ph.D., a senior consultant with the Tropical Science Center in Costa Rica. "Simply creating more protected parks is a negative solution; the pressures are mounting and mounting fast. Even in Costa Rica, with the best park system in the world, in ten years the economic pressures will become intense to cut down even the parks." And parks don't even have a chance in less-developed countries.

The only long-term solution, he says, is to devise a means "of using the tropical rainforests without destroying them." Hartshorn is experimenting with this concept in Peru's Palcazo Valley. Using very basic technology (oxen) to clear narrow strips of land (25 to 35 yards wide) through the rainforest, the natives then sell the wood and other products from the clearing.

Because the clearing is narrow, bordered by high trees on either side, seeds are quickly dispersed by bats and birds into the open soil. Sun pours through the clearing in the canopy, allowing vegetation to regenerate.

Schnell says that poor tropical countries simply don't have the financial resources to protect their forests. Exacerbating that poverty is the growing gap between the rich nations of the North and the poor nations of the South, due largely to gross inequities in trade arrangements whereby Northern multinational corporations and banks make huge profits at the expense of the Third World.

"In the long run, rainforests cannot be protected as long as this inequity between North and South continues," he says.

"We are expected to not develop our territory for the benefit of the world," says Costa Rica's Gamez. "Meanwhile, the developed countries consume 60 to 70 percent of the world's resources and aren't doing much about conservation even in their own countries."

Some countries, such as Sweden and Holland, have donated more to Costa Rican conservation than the United States, "the wealthiest nation in the world."

"When you compare assistance for conservation to military assistance in Latin America, it's ridiculous," says Gamez. "We're talking several thousand as opposed to billions of dollars." (Costa Rica is the only country in turbulent Central America with no military.)

One thing that must change if the rainforests are to be saved is the role of large lending institutions such as the World Bank. Environmentalists say it funds grandiose projects such as huge dams, highways through forests, and massive colonization schemes without the least thought for ecology.

Robert Goodland, Ph.D., of the World Bank, agrees. "We haven't been very good," he says. However, he maintains that his employer is changing. Several years ago he was one of a handful of environmental consultants at the bank. Now he is one of nearly 100. "There has been a quantum jump in the right direction, but it takes time for that change to have effect," he says.

Meanwhile, although the overall picture can appear grim, there is certainly hope. Some U.S. officials, for instance, are taking note. Sen. Tim Wirth (D-Colo.) has addressed the issue of tropical deforestation in Congress. "The world is rapidly losing its tropical forests for a short-term economic gain and losing long-term environmental benefits as well as benefits for national economies," he says. "The United States must take a leadership role...and halt this destruction."

Rep. Benjamin Gillman (R-N.Y.) recently proposed the Cultural Survival Act, which would enlist U.S. government support of indigenous rainforest tribes. The lobbying group Cultural Survival is advocating a bill that would open U.S. markets to products obtained from untampered rainforests (such as Brazil nuts) and close U.S. markets to products ob-

TROPICAL FORESTS

Rodrigo Gamez, adviser to Costa Rica's president. (Russell Wild)

tained from razed rainforest (such as mahogany and other South American hardwoods, and Central American beef).

One of the best hopes for the rainforests are "debt-swaps," the brainchild of the Smithsonian's Lovejoy. These are complex financial arrangements between U.S. banks that hold billions of dollars in IOUs from developing countries, the central banks of these countries, and conservation groups. Basically, the U.S. banks sell their (uncollectible) loans for a small percentage in cash. The loans are converted to local currency to purchase tracts of rainforest that are entrusted to the local conservation groups.

This means that for $1 contributed to an organization such as The Nature Conservancy, The World Wildlife Fund, or Conservation International, approximately $4 (in the case of Costa Rica) can be used to protect the forests.

The murder of Francisco Mendes Filho has helped bring the rainforest issue to the attention of many in the United States and other developed countries. Public awareness is a first step toward accomplishing what Mendes strived for during his life.

"The tropical rainforests can survive," says Costa Rica's Gamez, "but people in other countries will have to see it as a common goal."

Earth Care Action

Lovely Cheetahs, Meter-Saved

By Judy Howard

You've gotten used to talking boxes at the zoo—now it's time for preservation-promoting parking meters.

In an effort to raise money for the protection of habitats of endangered species, the San Francisco Bay Chapter of the American Association of Zoo Keepers has devised the Ecosystem Survival Plan (ESP), which urges people to "give your change to make a change" by dropping coins into an unlikely receptacle: old parking meters. Donated by the San Francisco Department of Public Works, the meters will be redesigned so that when money goes in, a picture of an animal such as a jaguar or an anteater will race across the window.

Five zoos (including San Francisco's) have agreed to install the meters in strategic places on their grounds, and 25 others have expressed interest. With 114 million people visiting zoos and animal parks every year, there's a lot of potential money for the kitty. "If every visitor put 50 cents in the meter instead of buying a bag of peanuts, we could purchase 4.8 million acres in the tropics each year," says Leslie Saul, one of the plan's initiators.

While many zoos have captive-breeding programs designed to help save endangered species, Saul and another ESP organizer, Norm Gershenz, see limitations to that approach.

"If you want to save the jaguar forever, you must save its habitat," Gershenz says. "And as a bonus, you save the millions of species that share the jaguar's home."

The zookeepers are putting the meter money into the Guanacaste National Park project in Costa Rica, where they have purchased 16 acres so far. The goal of the project, headed by biologist David Janzen, is to establish a 295-square-mile national park to preserve existing habitat and to restore damaged areas of this tropical forest. Because the ESP works through The Nature Conservancy and The World Wildlife Fund, it has no administrative costs: 100 percent of the meters' proceeds goes toward habitat preservation.

So what do your coins buy? Three hundred dollars buys $2^1/_2$ acres of tropical forest, on which you might find 200 orchids, 10,000 mushrooms, 200 frogs, 0.04 anteater, 0.001 jaguar, and a million ants.

Of course, if you're just not in the mood, you can always buy 600 bags of peanuts.

From Sierra *magazine, March/April 1989. Reprinted by permission.*

Earth Care Action

For Africa: A Fast-Growing Forest

By Nancy Herndon

Since the terrible drought of 1969, deforestation has been building to a crisis in Burkina Faso, a small West African nation barely south of the Sahara. In 1985, the year that Henri Ye left to work for a Ph.D. in forestry at Duke University, Durham, North Carolina, the crisis hit.

That year, civil servants in the capital, Ouagadougou, spent an average of 20 percent of their salaries on firewood, the region's main fuel. Village women walked as far as ten miles a day in the dusty countryside to gather wood for cooking fires. In the hot, arid north, the harmattan—a dry, dusty wind—whirled millions of tons of denuded soil into the advancing desert.

Ye, a forestry professor at the University of Ouagadougou, has been searching for the hardiest, fastest-growing tropical tree in the world.

Now he is returning with *Gliricidia sepium*.

Ye believes that the nitrogen-fixing *Gliricidia*, which grows 15 feet in five years and will sprout again from a stump, can play a major role in reforesting Burkina Faso. The silvery-barked hardwood, indigenous to Central America, can grow from both seeds and cuttings. It can withstand the eight-month drought that is seasonal in West Africa.

"It is almost a weed," says Ye, noting that the tree is nearly impossible to kill. Branches—even the whole tree—can be cut for firewood without harming the prolific root system.

Moreover, *Gliricidia* leaves are palatable for domestic animals. And its roots have nodules that fix nitrogen in the soil, replenishing exhausted farmland. In Costa Rica and Nigeria, the tree is interplanted with crops as a natural fertilizer.

Ye spent the last three years studying the properties of hundreds of tropical trees. Supported by a $22,000 grant from the Rockefeller Foundation, he will take the *Gliricidia* seeds from Nicaragua, Guatemala, Nigeria, and Kenya and develop a strain of the tree adapted to Burkina Faso's soil and climate.

An easygoing man who is fluent in English and French, as well as Bobo and Dioula, Ye has extensive plans for how the new strain of *Gliricidia* could be distributed in Burkina Faso. He envisions Agricultural Extension agents, foresters, and schoolchildren spreading seedlings and information.

The principal target of his outreach is farmers, who make up about 85 percent of the population. "Farmers can begin to use [the trees] almost immediately," he says. They can plant them as living fences to mark property boundaries, frequently in dispute in this once-nomadic society.

Within one year, the farmers can begin harvesting branches for firewood. Later they can grow the trees with maize and sorghum, fertilizing fields and protecting the soil from wind and rain erosion after crops are harvested.

Sandy, semiarid, and landlocked, Burkina Faso, about the size of Colorado, is home to some eight million people, who traditionally farmed land until the soil was depleted and then moved on to let an area recover.

But modern population pressures have destroyed farmers' long-standing balance with nature. Nomads in search of new lands have moved into the Sahel region in the North, cutting trees, overgrazing sparse grasses, and abetting the spread of the Sahara.

In the central part of the country, farmers have cleared more and more forest to gain cropland and firewood. Today the landscape is stripped of trees within a 50-mile radius of Ouagadougou.

Ye expects strong support for his projects from the Burkina Forest Service. "The whole government is interested in reaching farmers," he says. "We cannot plant enough to compensate for the cutting."

Reprinted by permission from the Christian Science Monitor, *May 11, 1989. Copyright © 1989 the Christian Science Publishing Company. All rights reserved.*

WILDERNESS

With Fate of the Forests at Stake, Power Saws and Arguments Echo

By Timothy Egan

Postscript: "It's Sort of Kill a Tree, Plant a Tree" 156
A Woman Who Guards a Wilderness 157
Prop 70: How Californians Won $776 Million to Protect Wildlife, the Coast, and Parklands 160

In the American rainforests, a misted expanse of coastal evergreens running from northern California through southeast Alaska, more of the tallest and oldest trees in the world are being cut from public land than ever before.

The cutting, spurred by profound changes in the timber industry, threatens the last stands of ancient trees in the United States, as well as the throngs of animals and birds that have dwelled in the forests for eons.

The virgin timberlands that once covered much of the country have shrunk to a prized strip in the Far West, most of it owned by the federal government. These last ancient forests are being logged at the rate of 60,000 acres a year—three times the annual harvest at the height of the post-war building boom.

Some tree experts think only two million acres, about 10 percent, is left of the forest that was here before colonial times. Other estimates are as high as six million acres, or 30 percent. Some scientists and conservationists think the old forests will be gone in three decades if logging continues at the present rate. Others say it could be 15 years.

From the New York Times, *March 20, 1989. Copyright © 1989 by the New York Times Company. Reprinted by permission.*

An Industry in Change

The cutting stems from industry changes as predictable as automation and as hard to control as foreign prosperity. Foreign timber companies are able to outbid American companies for logs from state and private lands. That has put more pressure on American companies to log the last remaining large source of raw wood for domestic use: the national forests. The companies plant new trees to replace the ones they cut, but seeds planted today will not become full-grown trees for decades.

For generations, the Forest Service was seen as the protector of the national forests, and the agency insists it still is. But some conservationists think the agency has become too cozy in recent years with the timber industry.

"Smokey the Bear has become Mr. Lumberjack," is how one conservationist group put it.

Fires Slow the Logging

Paradoxically, all this cutting has not brought prosperity to everyone. Although the huge timber companies that export heavily and rely on automation are doing well, there is double-digit unemployment in many small Northwest communities that depend on the small mills. Economists blame much of this on the export of raw logs, which go from forest to ship without touching a millworker's hand, and on automation. Never before have so many trees been cut in America by so few people.

Nationwide, the amount of timber stripped from the 156 national forests reached an all-time high of 12.7 billion board feet in 1987. That record would have been exceeded last year [1988] had not logging been curtailed by forest fires.

In the Northwest alone, a record 5.5 billion board feet (a typical 30-foot tree yields about 1,000 board feet) were logged from 19 national forests in 1988, about 20 percent more than ten years ago. Most of that was irreplaceable "old growth," virgin timber or trees two centuries or more old.

Before World War II, virtually no timber was cut from these forests, which President Theodore Roosevelt placed into the national trust in 1905. Surrounded by expanding cities and cleared-over private lands, the government forests have become the last home for some plants and wildlife at the very time they are being most intensively logged.

Conservationists say the 191-million-acre national forest system is in grave jeopardy. About 32 million of those acres have been designated "wilderness" and are inviolate. Twenty-three million acres of that land is in Alaska. In any event, much of the wilderness land is mountainous rather than timber-bearing.

About 10 percent of the national forest system is in the Northwest, and virtually all of the national system's virgin timber is there.

One Log in Four

By the year 2000, timber—when measured by actual volume of wood, not number of trees—will be cut nearly twice as fast on national forests as new trees can replace it, according to a Forest Service analysis done this year [1989].

One in four logs cut from the West Coast, including Alaska, was sent overseas last year, an all-time record, according to the Commerce Department. In the last ten years, logs equal to a forest of 600,000 acres were sent overseas.

Fine-grained Douglas fir, hemlock, spruce, and cedar, the most prized trees in the softwood lumber trade, come from the coastal rainforests, as do redwood and sequoia. Wood from these forests built the frontier towns of San Francisco, Portland, Seattle, and many of the houses constructed in that post-war boom.

Now the wood from those forests is the raw material for a Japanese home-building boom.

When Will They Disappear?

Small mill owners used to bid for wood from state and private forests. But in the midst of the export boom, small mills have been unable to match the bids of foreign log buyers, mainly the Japanese, who are buying up more and more logs from state and private lands.

As a result, the small mill owners are pressuring the Forest Service to let them cut more trees from federal land, the only area from which log exports are prohibited. Big companies are also pressuring for more access to national forests as they cut the last of their own trees.

At the rate of present cutting, most of the nation's remaining old forests will be gone in the next 30 years. "If you turned the forest-products industry loose on that land, it would all be gone in perhaps 20 years," said Dr. Jerry Franklin, a Forest Service scientist and a professor of forest ecosystems at the University of Washington.

The Wilderness Society has done studies that say that America's rainforest, with the trees and animals as they exist today, could disappear within 15 years.

The national forests of Washington and Oregon provide nearly half of all the trees cut from federal timberlands. Nurtured by an annual rainfall of up to 200 inches, the trees of the Northwest can grow up to 300 feet high and live for more than a thousand years.

Forest Service officials say they are setting aside some of this public land for future generations while trying to keep private timber companies supplied. But they are mired in controversy over how to manage these forests, and they are not even sure how much of the original American woodlands are left.

Bird's Fate Spurred a Count

In the mid-1980s, the Forest Service tried to calculate how much of the virgin forest was left. The survey was prompted by lawsuits seeking to protect the northern spotted owl, a rarely seen bird that nests exclusively in old forests of the Northwest.

The government hired Dr. Franklin, considered the nation's foremost expert on ancient forests. He and his committee defined original forest land as trees 200 or more years old. Contrary to a view held by some foresters and timber industrialists that some of these forests were so old they were actually decaying and biologically worthless, Dr. Franklin said the old timber stands made up an irreplaceable ecosystem.

Later, when Forest Service officials counted how much old growth was left, they did not use Dr. Franklin's definition. Relying largely on maps and aerial surveillance, and counting some lands that were already scheduled to be logged, the agency estimated that up to six million acres remained. When asked about this, Forest Service officials said using Dr. Franklin's definition would have been too difficult.

Now the Forest Service is doing a new inventory. Dr. Bruce Marcot, a Forest Service research scientist who is heading it, said the figures will probably be closer to Dr. Franklin's estimate that only two to three million acres of the ancient forest is left on public land, down from the service's earlier estimate of nearly six million acres.

Timber for Toyotas

Forest Service officials, who have set aside 5 percent of the old growth forest for the spotted owl, say they cannot reduce logging of the ancient trees by any significant level without causing havoc in the small, timber-dependent communities. But they say more of the rainforest could be saved if fewer trees that are now cut from state and private land were exported.

"There is no question log exports are affecting supply in the Pacific Northwest," said George Leonard, associate chief of the Forest Service. "But if we want to buy Sonys and Toyotas from Japan, we've got to sell them something they want."

Some big timber companies say overseas buyers are indeed paying prices too good to resist. Faced with pressures from Wall Street to increase returns on slow-growing timberlands, some of the big companies have accelerated their logging of land they planted, or have diversified into urban real estate after clearing their land.

There is a rising sentiment in the Northwest and on Capitol Hill to ban exports of raw logs, an idea embraced by conservationists and small mill owners, most of whom do not own any of their own timberland. According to a Forest Service study, 3.5 American mill jobs are lost with every million board feet of timber that is exported instead of being milled in the United States.

"Decks at Japanese mills are piled high as Mount Fuji with logs from the Northwest, while mills here at home are scraping for leftovers," said Rep. Peter DeFazio (D-Ore.). "We're facing the greatest timber supply crisis in our history, while Japanese mills are running around the clock."

Rep. DeFazio, whose district in Oregon cuts more timber than any other in the country, has introduced a bill that would allow states to ban raw-log exports. The bill is opposed by Rep. John Miller (R-Wash.), who says the states should not be able to restrict free trade.

Some scientists and government policymakers are calling for a new approach in forest management. Dr. Franklin, in a speech delivered at a forestry conference last month [February 1989], urged the Forest Service chief, F. Dale Robertson, and timber-company

---

Earth Care Action

Postscript: "It's Sort of Kill a Tree, Plant a Tree"

By Eric Malni

Everyone has heard about their old cars being ground up and recycled as brand-new Japanese automobiles. Well, this is about old trees being ground up and recycled as new foliage.

Last month [December 1988], the Los Angeles Bureau of Sanitation asked residents to drop their old Christmas trees off after the holidays at any of seven pickup points around the city so they could be put to some better use than buried trash.

According to Brent Lorscheider, assistant manager of the bureau's solid-waste task force, about 3,000 people complied.

The trees are being shredded into mulch and applied as a soil conditioner on flower beds, shrubbery, and other landscaping at the Lopez Canyon Landfill at the north end of the San Fernando Valley.

As a bonus, each person who dropped off a tree was given a seedling—a small, drought-resistant conifer—to plant as a sort of a replacement for the departed Christmas trees.

"It's sort of kill a tree, plant a tree," Lorscheider said with a chuckle. "It eases all our consciences somewhat."

Lorscheider said that since it was the first time the city had tried the program, "we had no idea how well it would work."

"I'd say it went pretty well," he said. "We got a little bit of everything, but for the most part, people did what they are supposed to and just dropped off Christmas trees.

"More than anything, it got people thinking about recycling," he said. "We're going to try it again next year."

From the Los Angeles Times, *January 17, 1989. Copyright © 1989 Los Angeles Times. Reprinted by permission.*

leaders and conservationists to stop viewing the last ancient forests as either one big wilderness or a tree farm.

What is needed, he said, is a management plan that will meet the needs of both nature and man. If such a new view is not adopted in the near future, he said, these last stands of forest will be gone.

Earth Care Action

A Woman Who Guards a Wilderness

By Nancy Herndon

Her gray hair swinging, Margaret Nygard climbs nimbly down a granite ledge slick with fallen leaves. Steadying herself with her dogwood walking stick, the elfin 64-year-old hikes briskly down the gorge.

"This is the crane-fly orchid," she says, pausing briefly to point out green and purple leaves that spring from the forest floor. "It blooms in the summer, tiny little orchid flowers, but the leaves die back then. If you don't know where to find it, you miss it."

By a clump of green she stops again. "There is trailing arbutus. It's rare, because its ecology requires a fungus." She leans forward on her stick, her round face crinkling with satisfaction. "We've been watching that patch for 25 years."

For longer than most of the nation's environmental protection laws have been written, Margaret Nygard has been watching over the wild and delicate beauty of the Eno River Valley, 40 miles of wilderness meandering through the rapidly urbanizing piedmont of North Carolina.

When she and a group of neighbors first protested the river damming and development of the valley back in 1965, local newspapers and businessmen called her a Rip Van Winkle, too out of step with modern times to understand that progress requires the destruction of wilderness.

Now they call her a visionary, a pragmatist who acted while there was still time, back when land prices were reasonable and nature lay pristine in big blocks that could be strung together into a major public reserve.

As the driving force behind a citizens' group called the Association for the Preservation of the Eno River Valley, Nygard helped design, promote, and fund the creation of the 2,028-acre Eno River State Park. The park links federal and local reserves to form a greenbelt 40 miles long.

The Eno River Association now includes 700 dues-paying members. They have raised $227,000 to buy parkland, negotiated creative compromises with developers and politicians, and for ten years have held an annual fundraising festival. The most recent one was attended by 35,000 citizens.

Last year [1988] the association received a national "Take Pride in America" award at the White House. This year they are launching a major land drive to bring 885 more acres—including rare plant habitat—into the park.

Most of these activities have been organized

from Nygard's kitchen, where the association maintains its telephone. She has been the group's executive vice president for 20 years.

"She is a role model for other groups," says Donald Reuter, public information officer with the North Carolina Division of Parks and Recreation, which nominated the association for last year's Take Pride award.

"She exhibits a genuine concern for the environment and somehow reaches the right people. She is an extremely valuable person to North Carolina," he says.

Soft-spoken and modest, Nygard seems an unlikely mastermind behind the actions of thousands of citizens. If not a Rip Van Winkle, she is clearly a person unbound by the conventions of this modern age.

With Holger, her husband of 45 years, she lives in a 175-year-old house in the woods above the river. She doesn't drive a car. Except in the dead of winter, she is usually barefoot.

But in her own unassuming way, she has been able to reach a wide range of citizens rarely united by more conventional leaders. "We didn't say 'Don't, don't, don't,'" she says of the association's efforts. "We said, 'Do, do, do.'"

Her ability to rally the business community to her cause netted her a $5,000 award last year from the Conservation Fund of Alexandria, Virginia, which annually recognizes an environmentalist who achieves significant conservation by cooperating with business groups.

A coalition with the Washington, D.C.-based Nature Conservancy first gave the Eno River Association tax-exempt status for landowners' donations.

By talking with elderly people and researching old letters, deeds, genealogies, and newspapers, the association was able to publish the river valley history and gain the support of the area's older families.

The group once persuaded some Green Berets to take citizens on raft trips down the river. Another time it organized National Guardsmen and high school students to move an old gristmill down from Virginia to re-create a historical site.

So many people have been so involved in helping create the park that Nygard now laughs off her leadership image as "myth-making." She routinely refers to her efforts as "us." But others credit her with providing the strategy behind all the diverse and disparate elements.

"Someplace in her head all of this synthesized into a whole," says Carol Charping, a member of the Eno River Association for 15 years. "It's Margaret who has the fine sense of purpose."

Her love of nature has been a lifelong passion, a fundamental bond she shares with Holger. As a young couple (they married when she was 19), the two cycled up Canada's precipitous Fraser River Canyon on bicycles that would horrify today's high-tech cyclists.

Her chosen field of study was English language and literature, in which she and Holger received doctorates from the University of Berkeley. But she was raising four children in the old house above the river when the city's plans to dam the river and flood the valley for a reservoir ignited her activism.

"There's something very stagnant about a man-made lake," she now muses. "To us a river is a very vibrant thing. Changing moods, changing seasons. It's kind of a restorative, if you go walking along the river. It replenishes after walking on pavement and being under city lights. It's very genuine."

Her family was swept up into Margaret's environmental work. Holger, a founding member of the Eno River Association, has worked beside her for more than two decades.

Margaret Nygard is the guiding force in preserving North Carolina's Eno River valley. (Photo courtesy of Union Camp Corp.)

His photographs are among those in the association's exhibitions of Eno River scenes displayed around the state.

Their oldest daughter, Jenny, is an artist whose work includes botanical drawings and posters for the annual Festival for the Eno. Their youngest son, Erik, is a state park ranger.

Nygard herself manages an astonishing array of activities. As vice chairman of the Durham County Open Space Committee, she is working to preserve other undeveloped lands for public hiking and recreation. Last summer she spoke before a U.S. House subcommittee on the importance of federal funding for conservation.

One recent week she spent several days lobbying against a new road that would fragment the greenbelt along the lower Eno. Another day she organized public pressure on the state Division of Environmental Management to stop a textile mill from discharging dyes that color the river.

"We're trying to work it out as nonconfrontationally as possible with industry and government," she says of the dye dispute, speaking with the gentle courtesy that is her hallmark. But her voice holds a note of unwavering commitment that has led some to describe her as relentless. Confrontation is not her style. But neither is surrender.

Her biggest task at the moment, she says, is to try to help fill in the state park's "missing links"—885 acres still privately owned and subject to development.

"Yesterday we bought outright a piece of property for $100,000," she says. She seems slightly in awe of the accomplishment, but she is already brainstorming about the possibilities.

The land, with the former Girl Scout camp next to it, has "the potential of holding a youth hostel and a ranger station, possibly an educational center," she says, noting that a new high school is planned nearby. "We've got to get the kids involved."

"I see the whole effort as unending," she explains. Then she picks up a soft-drink can from beside the river and hikes rapidly down the trail.

Reprinted with permission from the Christian Science Monitor, *March 7, 1989. Copyright © 1989 the Christian Science Publishing Company. All rights reserved.*

Earth Care Action

Prop 70: How Californians Won $776 Million to Protect Wildlife, the Coast, and Parklands

By Esther Feldman

In June 1988, Californians passed Proposition 70, a $776 million California Wildlife, Coastal and Parkland Conservation Act, which provides critical wildlife habitat, coastal and park lands throughout California.

Prop 70 has been described as the greatest environmental victory in California since passage of the 1972 Coastal Act and is the first park bond act in history to be placed on a ballot through the citizen's initiative process. Although this accomplishment is an impressive one, the diverse coalition that formed to support Prop 70 and to raise the funds necessary to pass it is at least as noteworthy: Prop 70 was supported by developers and business leaders, farm, civic, and conservation groups, Democrats and Republicans, sportsmen's and humane organizations, private land trusts and major corporations, and hundreds of individuals.

Times have changed. The outspoken contentiousness that marked the environmental movement of the 1960s and 1970s—and that was so necessary during those years—is no longer enough, or even effective. This is not to say that the environmental battles are not still here to be fought. But the realities of the 1980s call for a different sort of action, and it is increasingly vital that environmentalists work with their opposition to find solutions that are acceptable to both. Prop 70 strongly demonstrates that this kind of coalition building can work and that it can allow organizations and individuals with highly diverse viewpoints to reach common ground on controversial subjects.

The need for this kind of approach cannot be stressed strongly enough. The urban growth pressures that led to the necessity of creating Prop 70 continue to increase as we make our way into the 21st century. If we want environmental concerns to be seriously considered, we must deal with those concerns in the context of the economic realities of our society. The wide-ranging appeal of Prop 70 during the campaign was the measure's focus on preserving the quality of life in California and on protecting open space, wildlife habitat, and natural lands in order to enhance urban areas, tourism, and business development. We were able to show that preserving these lands made economic sense.

Prop 70 was conceived of by the Planning and Conservation League (PCL), a small statewide environmental lobbying organization devoted to promoting sound environmental laws. The effort began in 1984 as a PCL Foundation research project on the feasibility of urban greenbelts. We began with a volunteer intern, and the project quickly grew into a full-time one. The research led to the development of an urban greenbelt bill in 1986. PCL sponsored this legislation and worked to get it passed through the state legislature.

However, despite popular support, the legislature failed to act on this bill in 1986, along with two other land conservation measures that would have provided funds to preserve wildlife habitat and the coast. Led by Jerry Meral, PCL's executive director, we took these three measures and combined them to form the basis of Prop 70.

Why an Initiative?

We decided that the only way we would be able to get any significant amount of money for land acquisition would be to place a conservation measure on the state ballot through a citizen's initiative. All other funds for the conservation of wildlife habitat and parkland had been exhausted, and the need to preserve our wildlife, natural and park lands was growing more and more critical.

California's tremendous urban growth (at the rate of an entire San Francisco every year) seriously threatens the natural scenic and open-space lands that are an integral part of the quality of life in our state. Many of these forest, river, coastal, and ridge lands were in great danger of being lost forever unless we took some action.

Forming the Coalition

We brought together a coalition of statewide conservation groups to form Californians for Parks and Wildlife (CALPAW) to direct the initiative effort. CALPAW included PCL, the Sierra Club, National Audubon Society, Defenders of Wildlife, California Waterfowl Association, and experts in park planning, wildlife habitat preservation, and initiative campaigns. Two of PCL's staff became campaign director and assistant campaign director, and an office manager was hired as well. These three people were the core staff through the entire campaign.

Throughout the planning process and the actual writing of Prop 70, we worked to build a solid coalition of support that would result in seeing the measure passed. We met frequently with representatives from the building industry, farming interests, timber companies, and oil corporations as well as with developers, realtors, and key business leaders. We made an effort to consider the concerns of all these interests and to make sure that we were not including anything in the measure that any of these groups would really object to. We knew that widespread support of the measure would be critical to its passage. In June of 1987, CALPAW launched the signature drive to qualify Prop 70 for the June 1988 state ballot. More than 20,000 volunteers collected over 600,000 signatures to qualify the measure.

What Prop 70 Does

For the most part, Prop 70 follows the traditional format for California park bond acts.

Twenty thousand volunteers gathered more than 600,000 signatures and three-quarters of a million dollars to get Proposition 70 on the ballot. (Tracy Grubb)

(Bond acts are measures that allow the state to issue bonds for specific purposes such as parks, schools, jails, etc. Investors buy these bonds, which are paid back out of the state's general funds over 20 to 30 years.) It is unusual because it includes specific sums of money for specific projects throughout the state. Two-thirds of the $776 million provided will be spent to actually purchase land, and over 70 specific projects totaling nearly $300 million are written into the measure. This means that individual areas throughout the state are guaranteed protection, including such diverse areas as our magnificent Redwood State Parks, critical San Francisco Bay and Central Valley wetlands, key monarch butterfly habitats along the coast, and the threatened Santa Monica Mountains in Los Angeles, to name but a few. Funds are also provided to restore and enhance trout and salmon streams, plant trees in urban areas, preserve coastal areas, maintain productive farmlands in agricultural use, and protect critical wildlife habitat for rare and endangered species.

How It Was Done

None of the sponsoring organizations had the funds necessary to run a statewide initiative, so we knew we would have to raise all of the funds for our estimated $750,000 budget if we wanted to get our measure on the ballot at all. (This budget reflects the size of California.) We asked groups that wanted us to include specific projects to make pledges of either signatures or funds. We received pledges of over 600,000 signatures and over $650,000—and we collected all the signatures and nearly all the funds during the course of the campaign—probably an all-time record of collecting on potential pledges.

These donations ultimately enabled us to meet our budget and to gather the signatures we needed. The groups that made the pledges also formed the core group of our support that lasted through the entire campaign. People were much more likely to support Prop 70 because they could see how it directly benefited them and their local area. These individual projects allowed us to gain the support of local businessmen, developers, chambers of commerce, farming groups, conservative political leaders, and many, many others.

We hired a statewide coordinator (Ken Masterton) to oversee the entire signature campaign, as well as nine regional staff people who worked in local areas throughout the state to coordinate local volunteer efforts. These coordinators organized the more than 60 different local groups who had pledged to collect signatures. Because people at a local level really cared about projects in their area, they worked very hard to make sure that Prop 70 would make it onto the ballot. These thousands of supporters became ready-made grass roots support during the election campaign.

Opposition

We were extremely careful not to include any projects that were at all controversial, because we knew we could not afford to fight any organized opposition. We planned on, and carried out, a very low-budget campaign. We were able to do this only because our opposition was minimal. Farm and cattlemen's groups and the State Chamber of Commerce opposed us, mostly because they are opposed to buying any additional lands for any park and recreational purposes.

Implementation of Prop 70

Now that Prop 70 has passed, the Planning and Conservation League Foundation is working to ensure that it is properly implemented by all the state agencies who are in charge of the funds. The foundation is tracking all the Prop 70 programs, monitoring the bond process, and keeping a detailed record of all land acquisitions and program accomplishments.

In an era where we continue to fight environmental battles and seem to expend huge amounts of effort to defend what we have so dearly won, Prop 70 stands out as an *active* step by the people of California to help solve one of the most pressing environmental problems in the state: that of disappearing wildlife habitat, park and coastal lands, and open-space areas. The ramifications of Prop 70 are great. Not only will thousands of acres of natural lands be protected forever, but the coalition of support also paves the way for future alliances that will continue to work together to protect our magnificent and irreplaceable natural resources—forever.

You Can Do It, Too

The PCL Foundation has also published a comprehensive manual on how to design a measure such as Prop 70. The publication includes a detailed breakdown of:

- Initial research and background information needed
- Organization of the entire effort
- How to write such a measure
- Fundraising
- Endorsements
- Signature-gathering campaign
- Election campaign

The manual speaks primarily of the experience we have had in designing a statewide measure, but it certainly can be applied to local or regional areas as well. To receive a copy of the manual, or more information on Prop 70 or PCL, contact PCL at 909 12th Street, Suite 203, Sacramento, CA 95814, (916) 444-8726. There is a $10 charge for the Prop 70 manual; payment should be made to the Planning and Conservation League.

From the Whole Earth Review *(27 Gate Five Road, Sausalito, CA 94965), Spring 1989. Reprinted by permission.*

WILDLIFE

Showdown at Sea

By Russell Wild

A World without Whales? 168
The Return of the Giant Clam 171
Turtle Rescue 175
Drought and Wetlands Drainage Take a Heavy Toll on Many Species 176
People Who Make a Difference 177

Our global garden is in trouble. Our rainforests, coral reefs, and wetlands are rapidly vanishing. And with them go their inhabitants—plants and animals, many of which scientists haven't had a chance to classify. Humanity is driving at least one species a day into extinction, and by the end of the century, biologists expect that rate to approach one species *per hour*.

"A crisis situation" is how Jack Sites, Ph.D., associate professor of zoology at Brigham Young University, describes it. "There's a web of life that we are destroying, and people have yet to realize the interdependency of the species." To grasp that notion of interdependence, consider that as many as half the drugs used in modern medicine had their origins in wild plants and animals.

Even those endangered species that have stirred the most public concern—the California condor, the mountain gorilla, the rhinoceros, and others—are still greatly imperiled. And nothing illustrates that more dramatically than the case of the vanishing whale.

Indeed, the Save-the-Whale movement that flourished in the 1970s—and which by 1986 had sparked an international

From Organic Gardening, *September 1988. Reprinted by permission.*

treaty banning the hunt of the great mammals—has miles to go before it sleeps. Such nations as Japan, Iceland, and Norway have continued to hunt whales, exploiting loopholes in the treaty. Unless they stop, tens of thousands of whales will be slaughtered over the course of the next decade.

The eight largest of the great whales are all listed as "endangered" species under the U.S. Endangered Species Act. Biologists agree almost unanimously that certain of these species, such as the blue, right, and bowhead whales—at probably less than 3 percent of their preexploitation numbers and scattered about the globe—live a precarious existence, and would even in the absence of a hunt.

Others, such as the sperm, sei, and humpback whales, can probably recover—but only if protected. The gray whale, at least the Pacific gray whale (for the Atlantic stock has long since been harpooned into oblivion), has, thanks to U.S. protection efforts, already made an astonishing comeback. But past protection efforts aren't enough, although the Save-the-Whale movement has made remarkable progress since the 1960s, when more than a dozen nations combed the seas, killing an estimated 60,000 whales a year. In those days, whaling had taken on a new dimension, far different from the romantic era when Herman Melville's Captain Ahab sought Moby Dick with handheld harpoons in his three-masted whaler, the *Pequod*.

By the 1960s, whalers were firing massive harpoon cannons from swift, diesel-powered, killer ships armed with radar and sonar. Accompanying the killer ships were multistoried, 600-foot-long, steel-hulled monster factory ships, capable of processing an entire adult whale in as little as half an hour.

Perhaps the only aspect of whaling that hadn't changed was the cruelty of the hunt—for as massive an animal as the whale cannot be killed humanely. Regardless of whether a harpoon is thrown by hand or shot from a cannon, the warm-blooded, air-breathing animals can agonize for hours before death. Often several harpoons must be fired.

Yet despite the brutality, and despite clear signs that several whale species were rapidly vanishing, the world of two decades ago remained dully apathetic.

But the public's consciousness was eventually pricked when scientists reported that the magnificent blue whales, the largest animals ever to inhabit the earth, were rapidly disappearing. The giant mammals got their first burst of human support at the 1972 U.N. conference on the Environment, when 52 nations voted unanimously for a resolution to adopt a ten-year moratorium on whaling.

The passing of the U.S. Marine Mammal Protection Act later that year, and the U.S. Endangered Species Act in 1973, served to protect whales in U.S. waters and to close U.S. markets to whale products.

In 1975, a fledgling environmental group called Greenpeace redefined civil disobedience. With all the drama of film epic, and using a blend of tactics taken from Mahatma Gandhi and Muhammad Ali, Greenpeace volunteers launched portable rubber rafts with outboard engines in the direction of the whalers. Like little buzzing Davids, they intercepted the Goliath Soviet, Japanese, Spanish, and Icelandic whaling fleets, and with a maternal-like sense of protectiveness, positioned themselves between the whalers and the whales.

By the late 1970s, the number of whales being killed had declined dramatically. The decline was partially attributable to the difficulty hunters were encountering in finding the remaining whales. The decline was also due, in large part, to televised images of Greenpeace's splashy encounters with the whalers, scenes beamed into millions of homes worldwide.

The public was moved. With organizing efforts by groups like the International Society for the Protection of Animals, Project Jonah, and Friends of the Earth, demonstrations in the United States, Canada, Australia, and Western Europe ensued, forcing country after country to bow out of the whaling business.

By the early 1980s, things were turning from bloody to downright rosy for the whales. The International Whaling Commission (IWC), at the urgent behest of its scientific committee, was for perhaps the first time in its almost 40-year history putting the conservation of whales before the profits of whale meat. No political mystery here—the whaling nations that once formed a majority of voting interests in the IWC had steadily boiled down to a small and decidedly unpopular minority.

The IWC voted in 1982 for an indefinite moratorium on all commercial whaling, to begin in 1986. The vote was 25 nations for, 7 against—an overwhelming victory for the whales. Protecting them was now not only the will of the people, it was international policy as well.

But even before the moratorium started, it began to unravel. Three nations—Japan, the Soviet Union, and Norway—pooh-poohed the whole idea, ignored the IWC, and went on their merry hunting ways. Iceland and South Korea, meanwhile, found what one prominent U.S. conservationist calls "a crummy little loophole," which has since been pushed open wider and wider to the point that the IWC has come to appear shamelessly feeble.

The major pusher has been, and continues to be, Japan. Last year, the nation agreed in principle to abide by the moratorium. In March 1987, with the greatest pomp and circumstance, its whaling vessels having returned from their annual Antarctic cruise, it wistfully announced to the world press the end of an era of Japanese commercial whaling. But this turned out to be the end of nothing.

This year [1988], those same ships, using the same harpoons, are out hunting the same icy waters. Whale meat in Japan is a delicacy, bringing enormous prices in fancy downtown restaurants. Most of what is not turned into snacks for humans is ground up for pet food. When the whalers return home, by their own admission, they will be selling the same whale meat to the same merchants in the same markets as before. There is a difference, though— now the Japanese say they're doing it for "science."

Here's the loophole: Under international treaty, a nation has the right to kill whales if it's done for science. And that's just what Japan says it's doing. And what it says it will keep on doing—to the tune of more than 10,000 sperm and minke whales over the next 12 years.

Japanese "research whaling," needless to say, means a lot of spilled blood on the oceans. It also means a lot of boiling blood on land, among environmentalists, animal rights advocates, conservation-minded politicians, and perhaps most of all, scientists. To the vast majority outside Japan, science is being seriously perverted.

"This is nothing but a scam," says Roger Payne, Ph.D., senior scientist at the World Wildlife Fund, president of the Long Term Research Institute, and possibly the best-known whale researcher in North America. (It was Payne who first discovered that humpback whales sing long, complex songs, and that blue and fin whales have "voices" that may travel underwater for hundreds of miles.)

"I'd much prefer that the Japanese just said, 'To hell with you all—we're going to keep on whaling'; that, I would understand," says Payne. "But to pretend that they're doing this for science, it's hard to convey just how frustrated, irritated, and annoyed that makes me."

His outrage is not uncommon, particularly among his fellow members on the IWC scientific committee, who voted overwhelmingly this year to condemn Japanese "scientific" whaling as wholly without scientific merit. In fact, only the scientists from Iceland and Norway offered much support for their Japanese counterparts. Here again, no political mystery: Iceland and Norway are currently pursuing their own "scientific" whaling projects—and selling most of the meat to Japan.

"Totally irrelevant," says Alan Macnow, long-time U.S. spokesperson for the Japan Whaling Association. Neither politics nor profit plays any role whatsoever in "research" whaling, he says; "it's strictly for science." To know population trends among the whales, Macnow maintains, you have to know their age distribution, mortality rate, and birth rate.

To discover these things, Macnow says, whales have to be cut up. He says studies of the teeth and the buildup of earwax determine a whale's age. A good look at a whale's belly determines its gender. And internal probing determines whether a female is pregnant. "You can't just parade whales in front of a calculator to find out these things," he says.

"If this were really an example of Japanese science, the Sony Walkman would be pushed around in a wheelbarrow," says Payne. Even if Japanese were genuinely concerned about whale populations, he contends, "they wouldn't be killing them in droves...they wouldn't have to. There are alternative means of gathering information on whales."

Michael Tillman, Ph.D., chief of the conservation science division at the National Marine Fisheries Service, and another member of the IWC scientific committee (he is a former chairman), wholeheartedly concurs. The Japanese, he says, "are using the research loophole to conduct business as usual." New nonlethal techniques for sizing up whale populations (such as "genetic fingerprinting" through the

---

A World without Whales?
By Russell Wild

When not agonizing over how to stop the whalers, some conservationists and scientists are beginning to wonder what a world without whales might be like.

More specifically, some scientists believe that the whales may play a vital role in nature. It is possible, for example, that sperm whales are to the deep waters they inhabit what earthworms are to a garden. Sperm whales feed on large squid and fish in deep water and rest at the surface between long dives, doing much of their defecating at the surface. The tons of organic material that they defecate may be an important source of nutrients to the water's upper level, supporting the growth of planktonic algae, the first step in the massive food chain.

Similarly, another theory has it that the Pacific gray whales, with their hangar-sized jaws and lust for subterranean worms, may scoop up and send into suspension as much mud in one year as does the Mississippi River—and therefore could play an integral role in the shaping of American's continental shelf. In addition, the grooves they leave on the ocean bed are thought to serve as a primary habitat for many animal species.

From Organic Gardening, *September 1988.*

use of skin samples taken with darts) could potentially yield much better data than anything the Japanese may garner from dead whales, he says.

But Macnow alleges that Tillman and Payne, as well as many other representatives of the IWC, have been lobbied by environmentalists to the point that they can no longer see the ocean for the waves. Macnow contends that only the Japanese scientists can be wholly objective because of mistaken beliefs in America and elsewhere that whales are intelligent and equally mistaken beliefs that they are endangered.

Now, if disagreement over Japan's "scientific" whaling program weren't hot enough, contentions that the world's largest mammals may be no smarter than big fish—and not endangered—is enough to infuriate many scientists. Driving home this point, Macnow says, "If whales were so intelligent, you'd think they'd understand the nature of tides. Why then do they beach themselves so often?"

Payne disagrees. Studies of dolphins, the smallest members of the whale family, show that "their brains are at least as complex as a human being's—maybe more," says Payne.

Lou Herman, Ph.D., director of Kewalo Basin Marine Mammal Laboratory at the University of Hawaii, says Macnow's example of whales' beaching is fairly ironic. Sick whales often seek shallow water to be able to rest on the sand while keeping their blowholes above water. They also escape deep-water sharks at the same time. They do risk getting whisked ashore, and sometimes this happens, but swimming into shallow waters could still be deemed intelligent, not stupid, behavior.

Struggling Species

The Japanese argue that whale stocks are hardier than Western scientists believe. In addition, they defend their whaling industry by pointing out that most of the whales being killed are minke whales, not an endangered species. Here—although counting whales is always an inexact science—the Japanese are probably right. Technically right.

The minkes are not endangered. They are the smallest of the great whales, and back in the heyday of the hunt no self-respecting whaler would have wasted his time with such a scrawny catch. Back when 100-foot, 160-ton blue whales, 70-ton right whales, and 50-ton sperm whales were plentiful, it made little economic sense to chase the "mere" 30-foot, 10-ton minkes. Only now that all the others have been driven into "commercial extinction" do minkes matter to the whalers. They are the leftovers.

Scientists fear the past is repeating itself. The first whale to suffer from man's interest was the right whale—so called because its sluggishness, hefty size, and tendency to float after it died made it the "right" whale to hunt. Then, as hunting only one species failed to prove profitable, the whalers turned to the next easier-to-find species—the sperms, the humpbacks, the fins, the seis.

"The whalers pretty much started with the largest and worked their way down, until every whale they ever hunted became endangered," says Payne. What about the minkes? "The minkes will go the way of the others."

Not that these "others" are safe from the hunt—far from it. The Japanese "research" plan calls for taking 50 endangered sperm whales a year. Iceland's plan, which has received no less flak from conservationists or whale scientists than has Japan's, will take 10 endangered sei whales and 68 endangered fin whales, in addition to 80 minkes this year. Norway's likewise controversial "research" calls for killing 35 minkes this year, and probably more in following years, from the incontrovertibly severely depleted Northeast Atlantic stock.

Should the heat to end "research" whaling get a little too hot for the countries involved or should the current number of whales being killed not be enough to fill their appetites, research of another kind is currently under way: to find new loopholes. According to Payne, the Japanese are particularly interested in an IWC loophole that allows for "subsistence" and "aboriginal" hunts. Keep in mind that current whaling activities in Japan are being done by major conglomerates, and that whale meat fulfills less than 1 percent of the population's protein needs.

Halting the Hunt
All this killing has Greenpeace and other environmental groups, not to mention the IWC, groping for answers. Tillman admits that if the IWC cannot soon get its three recalcitrant members to behave, other nations such as South Korea and the Soviet Union, which only very recently and very reluctantly gave up whaling, may jump back into the profitable games.

The key to halting the hunt, sustaining the IWC, and saving the whales lies with Japan, says Payne. "Japan is the market for the meat—without that market, Iceland and Norway would stop their shenanigans immediately."

Stopping Japan will be difficult, says Payne, not so much because the Japanese are hooked on whale meat, but because their culture still sees the whales as big, tasty fish, and because the whole matter has become an issue of national pride.

Meanwhile, some elected officials in Washington are starting to agree with the conservationists who've said for years that the U.S. and other governments should be doing something to end the killing. Rep. Don Bonker (D-Wash.), introduced a congressional resolution in January calling for the nations of the world to impose economic sanctions against any nation that continues to whale in defiance of the IWC.

"This brutal hunt is just a poorly disguised effort by Japan to continue its commercial whaling operations," says Bonker.

The U.S. government this past April announced that Japan will be barred from fishing in U.S. waters until it stops whaling. Several years ago (when environmentalists and their friends in Congress were lobbying the Reagan administration to act), such a move would have cost the Japanese millions of dollars. Today, it is largely a hollow gesture: Japan isn't fishing U.S. waters this year, as a result of recent U.S. efforts to protect its home fishing industry.

Groups like Greenpeace, The World Wildlife Fund, and the Humane Society are rallying support for the battles to come. Greenpeace is sponsoring an economic boycott aimed at Iceland and considering further action on the high seas. One group, the Sea Shepherd Society, went so far as to sink two Icelandic whaling ships in harbor in November 1986.

But whale defenders know they're fighting an uphill and, some fear, possibly a losing battle. "The Japanese position is so extreme and totally unreasonable, so flawlessly and seamlessly unreasonable, that it's hard to fight," says Payne.

Perhaps the onslaught against the whales will end with growing consciousness from within the three whaling countries. Greenpeace Director Steve Sawyer points hopefully to a growing ecology movement in Japan. And last year, 21 top Icelandic biologists called on their government to suspend the kill. Perhaps it will end as a result of economic and political pressure from outside.

The whales are waiting.

Earth Care Action

The Return of the Giant Clam

By Stephanie Ocko

Taufa'ahau Tupou IV, monarch of the last surviving kingdom of the Pacific, smiles majestically into a video camera and says, "You're bound to win with aquaculture." Heir to a 1,000-year-old tradition of tribal chieftains, the king sets the cultural standards in the archipelago of Tonga, located south of Fuji and Samoa. Chronically overweight, he is a vigorous proponent of physical fitness. No less is he aware of his country's need for environmental repair. On this day, he is speaking out on behalf of giant clams, one of Tonga's most impressive and most endangered species, for a video produced and directed by marine biologist Dr. Richard Chesher. "I think that the culture of giant clams should be encouraged, and where giant-clam circles are arranged, they shouldn't be raided by irresponsible people," the king rumbles. "If we succeed with giant clams, then we can go on to cultivate other species."

"The Giant Clam Circles of Vava'u," the 20-minute video that will include the king's endorsement, represents the first major step in Richard Chesher's campaign to spread a simple but important fact around Tonga: giant clams—the spectacular native beasts that grow up to a meter across—have been harvested nearly to extinction.

To a scientist such as Chesher, a 20-year veteran of Pacific marine biology, the clams' plight is obvious. But convincing Tongans of it is another matter. Chesher's idea is simple: If villagers and fishermen gather giant clams, arrange them in circles in shallow water, and leave them to breed, the clam population could replenish itself within a few years. He chose circles both because they would signal human involvement and because they would promote clam breeding. But though his intentions are excellent, Chesher is a *papalangi,* a white man, and therefore an outsider, and Tongans are slow to accept his advice. With the king speaking in behalf of the circles, however, Chesher believes he may finally reach his audience.

It is only a beginning. A modern cash economy has driven fishermen from throughout the South Pacific to fish for profit rather than simply to feed their families or villages. Giant clams have been one of the first victims of the cultural clash: In the last 20 years their meat has been frozen for export; for many more years, their shells have been sold for baptismal fonts or birdbaths. At the same time, the coral reefs where many of the clams live have been decimated by pollutants and destructive fishing methods.

Chesher believes that successful clam circles could demonstrate to the Tongans how they can begin to reverse environmental neglect and purposely nurture other endangered species. "The concept of leaving some large marine creatures alive in a protected area to have young and replenish the reefs is the very essence of conservation," he explains. "If the giant-clam circles can become a part of the island's lifestyle—a cultural tradition—we will have created a base for a more general environmental awareness." Ideally, environmental

Vulnerable to any poacher or hungry fisherman, the giant clams of Tonga have been harvested to near extinction. (Richard Chesher/*Earthwatch*)

awareness in Tonga will spread to other islands, where local governments can institute similar conservation measures.

For Richard Chesher, the first sign that all was not as idyllic as it seemed in the Pacific came in 1969, on a Pacific-wide survey sponsored by the U.S. Department of the Interior. Millions of crown-of-thorns starfish, Chesher found, were rapidly destroying coral reefs. "I realized then," he says, "that man's activities on the coral reefs had unexpected and long-range consequences." The use of pesticides and dynamite and careless reef fishing all combined to kill the living coral. The starfish, meanwhile, were forced to live in diminishing quarters on the remaining coral, creating an unprecedented opportunity for breeding. The resulting horde of starfish gradually overwhelmed the remaining reef, feeding on coral polyps.

Chesher thinks that the underlying causes of this environmental upheaval are "a lack of understanding, a lack of consideration." Population increases, the development of commercial fishing, government loans that allow fishermen to buy outboard motors for their boats, and extensive land clearing all contribute to disturbances in the delicate marine ecosystems of the coral reefs. In Tonga, women and children once walked the reefs barefoot at low tide, looking for shellfish; today people stride across the reef in shoes, prying off clams with iron poles, bush knives, and hammers. Fishermen used to catch octopuses with a lure that resembled a rat; today they scoop them out of their nests at low tide, destroying eggs as well as nests. Chesher's Earthwatch interviewers discovered in 1985 that Tongans did not believe that coral was a living organism.

The effects of the destruction are as varied as the causes. Broken coral can become infected with algae that act like a deadly virus. The algae are spread to healthy coral on the microscopic leg hairs of small creatures like crabs that inhabit the coral. So fragile is the reef that stress from radical temperature changes or pesticides can cause breakage that results in sudden coral death, called a shutdown-reaction. During a survey of 100 reef sites in Vava'u with Earthwatch volunteers in 1984, Chesher estimated that 65 percent of the reef was reduced to rubble.

For three seasons, a total of more than 150 Earthwatch volunteers have worked with Richard Chesher to reverse this trend. Today, the teams are trying to create clam gardens off the village of Falevai in Vava'u, in the most northern island group of the archipelago.

The biology behind the clam circle is fairly straightforward. Giant clams are all born males, but at the age of six years they become females as well. Once mature, they procreate by shooting out clouds of sperm, waiting about 20 minutes, then shooting out millions of eggs. When one starts, the group follows suit. Mating near each other, Chesher says, ensures that there will be a certain generational vigor: Depending on currents, the sperm of one clam may meet up with the eggs of another. The trouble is—because there are only a handful of giant clams left on Tonga—opportunities for group sex are increasingly rare.

For Chesher, the clams' plight offers an excellent opportunity to transfer the concept of animal husbandry from the land to the sea. The basic idea is very simple. Because the largest clams produce the most eggs—a 40-to-50-year-old makes 400,000 times more eggs than a clam at first female maturity—Chesher's teams in Vava'u are trying to convince the Tongan people to form circles of 100 big clams in shallow water. These circles will create ideal breeding conditions.

Tridacna derasa, or *Tokanoa mole-mole* in the Tongans' native tongue, averages a hefty half-meter across and is the largest of the island's giant clams. It is also the easiest to harvest. Unlike other species, it doesn't anchor itself to the reef and is easily lifted away by fishermen. And while the Tongan Fisheries Act of 1924, designed to protect fish in weirs, has been extended to protect clams when they are arranged in circles, Chesher believes the law is ineffective because it does not have the community's support. Only community pressure will hold off a Tongan fisherman or poacher.

Though giant-clam circles have been successful on other Pacific islands—on Palau, for instance, a breeding stock of 10,000 clams in a commercial aquaculture center spawned 100,000 young clams in 1984—convincing Tongans of their worth has been a tricky business. To a fisherman, creating a clam circle amounts to laying off fishing for profit and taking up nonprofit fish farming for three to four years. Many fishermen find it hard to bank on a distant payoff when they are struggling just to make enough money to feed and clothe their families.

How best to publicize the fact that clam circles are not only environmentally necessary but profitable in the long term? While working in Samoa, Chesher successfully made and distributed educational videos that described environmental problems and solutions. But the system was already in place: American Samoa has had a vigorous educational television program for both adults and children since the mid-1960s. Tonga is a different matter altogether.

Although English is the Tongans' second language and the islands have a 95 percent literacy rate, the print media are dicey. The *Tongan Chronicle,* the weekly government newspaper, sometimes doesn't even reach Vava'u, which has a population of about 10,000, plus a constantly changing group of ocean-going yachtsmen who put into the harbor from May to December, and scattered American, Canadian, and Australian Peace Corps volunteers.

Chesher theorizes that the real network of daily, usable, valuable information among Tongans is gossip. Every day at noon and all day on Saturday, Vava'uan men gather to talk on the steps of one of the department stores that line the main street. They are cavernous, old, colonial-style buildings with creaky wooden floors, glass display cases, and shelves filled with the latest from the weekly cargo ship—automobile tires from New Zealand, peanut butter and Coca-Cola from the United States. In the open-air farmers' market, women in long dresses chat and barter for vegetables as their children hide smiles from strangers behind their mothers' skirts. And on Sundays, everyone but the infirm goes to one of the many churches, whose rich choral music swells in the still morning air.

The countryside is sprinkled with neat houses surrounded by gardens, ubiquitous pigs, and cackling chickens. But at night, the interiors of most of the houses glow with blue TV light. The Tongans' two most appreciated appliances are sewing machines and VCRs, and the most colorful shop in the town of Neiafu, the capital of Vava'u, is the red and white video store.

In such an environment, Chesher decided the best way to spread information about the clam circles is through videos, which will lead, he hopes, to gossip. Working with Earthwatch volunteers, he put together a video script in which the district commissioner of Falevai, representing the community, cares for a circle by watching over the clams and periodically checking their health. One day he catches a poacher, whom he then educates about the long-term effect of the circle. Though King Taufa'ahau Tupou was never part of the original script, it was only a matter of time before he became involved in the project. After all, it was the king who, in 1986, officiated at the placing of the first Tongan clam circle, when fishermen gathered 100 *Tridacna derasae* and put them in the harbor in front of his Victorian palace.

When Taufa'ahau Tupou uttered the words, "You're bound to win with aquaculture," at the end of the video, it was a historic occasion: the first time any king had ever spoken in behalf of giant clams. Knowing for certain that the clam circles will work will make it easier for Chesher to sell another idea; a clam genebank preserve. The other challenge, to measure how well he might have instilled environmental awareness and awakened Tongans to the concept of caring for the clams and the reef—and spreading the conservation measures to other islands—will be the subjects of future research.

But for the moment, giant clams have hit the media: "The Giant Clam Circles of Vava'u" is available now in the video store in Vava'u, in schools, and in video rental parlors throughout Tonga, and it is scheduled soon for Tongan cable TV.

From Earthwatch, *June 1989. Reprinted by permission.*

Earth Care Action

Turtle Rescue

By Mindy Leaf

Endangered sea turtles will get a second chance at winning the race for survival. Concerted efforts by conservationists and Congress have resulted in the passing of a Turtle Excluder Device (TED) law, which will require shrimp fisherman to use TED nets when fishing offshore, beginning May 1 of this year [1989].

After much deliberation, the late-season legislation was finally approved as part of the Endangered Species Act. "All branches of the federal government have now looked at the evidence on the drowning of sea turtles in shrimp nets, and all have agreed that shrimp fishermen must begin using TEDs if sea turtles are to be saved from extinction," comments Marydele Donnelly, director of the Sea Turtle Rescue Fund at Washington's Center for Marine Conservation.

The nets are designed to allow 97 percent of captured sea turtles (which are air breathers) to swim free. The nets also provide an escape route for finfish and other marine species. With TEDs, everyone wins. At a cost of only $400 each, TEDs reward shrimpers with far more time-efficient, fuel-efficient, waste-free catches.

Previously shrimp trawls (the large standard nets) accounted for the "accidental" capture of about 48,000 sea turtles each year. So it's hardly surprising that the Center for Marine Conservation has hailed the new legislation as a "tremendous victory." Still, says Donnelly, "much work remains for all of us" in educating fishermen.

From Omni, *April 1989. Reprinted by permission.*

Drought and Wetlands Drainage Take a Heavy Toll on Many Species

The plight of wildlife in many areas of the United States worsened dramatically during the drought of 1988, but most conservationists were more concerned about the long-term effects of human activity on habitat than the harsh weather.

Ducks and other waterfowl were the hardest hit. For decades, the birds have struggled to maintain their populations despite the continual encroachment by people into their wetland habitats. Half the ducks in North America depend on the five million lakes and ponds in the north-central Plains States and southern Canada—the prairie potholes—as breeding grounds. Last summer [1988], fully 40 percent of the potholes evaporated under drought conditions—a situation that only compounded the real threat to North American waterfowl: Throughout the country, wetlands drainage for agriculture and other kinds of development continues at a rate of 300,000 to 500,000 acres per year.

After recording a modest increase in duck numbers in 1987, the U.S. Fish and Wildlife Service predicted that 1988 would see the second lowest North American duck migration on record—an estimated 66 million birds. "It's the worst time in the history of waterfowl other than the 1930s," observed the agency's director, Frank Dunkle.

Receding waters across the country due to the lack of rainfall, combined with excessive use for agriculture in some areas, also meant reduction of many fish populations. Fisheries experts do not expect full recovery of some western bass and walleye populations for five to seven years; it may take even longer for some trout fisheries in Montana to recover from the effects of heavy irrigation drainage.

Even major rivers became dangerously low and warm—thus inhibiting their oxygen-carrying capacity and threatening large numbers of salmon, as well as trout and bass. The situation became so serious in California's Sacramento River that the state released from Shasta Dam water representing $2 million worth of hydroelectric power to cool the river and save a generation of salmon fingerlings.

While waterfowl and fish appeared to decline the most in 1988, other kinds of wildlife also suffered from the dry conditions. The mast crop—the acorns and other nuts that are the principal food of many forest dwellers—was rated poor last year in the Midwest and South, where heat and lack of rain reduced the production of most nut trees and killed many of the older ones. Any major reduction of mast means a famine for deer, squirrels, and other nut-eaters—and consequently a hard winter for foxes, owls, and other predators.

Despite the hardships, however, scientists see the drought as a frequent, natural event. As National Wildlife Federation biologist Douglas Inkley put it, "We will certainly have starving animals and lower populations of some species this year, but I don't see any species going extinct because of this drought. The real problem is the continuing loss of habitat."

Last October, in an important victory for conservationists, President Reagan finally signed the reauthorized law, which will breathe new life into the nation's underfunded endangered animal programs. Among other things, the strengthened Endangered Species

Act now provides funds for monitoring the status of many declining species not yet officially considered threatened or endangered.

Meanwhile, wildlife did enjoy at least a few other clear-cut victories in 1988. Last July, for instance, the Environmental Protection Agency (EPA) indicated it would reject a controversial plan to build a recreation area at Lake Alma in southern Georgia. "The project would have wasted tax dollars on an economically questionable venture and would have ruined important wetlands and other habitat in the process," observed National Wildlife Federation president Jay D. Hair. Earlier in the year, the EPA had made it clear it was going to take a more active role—based on its new powers under the Clean Water Act—in protecting the nation's wetlands, by fining three developers thousands of dollars for illegally filling marshlands in Illinois.

And despite continued habitat loss, there were other occasions for quiet celebration: reports of preliminary successes in the program to reintroduce the red wolf to the wild in North Carolina; news that the once-endangered brown pelican is not only flourishing but has extended its range in many areas; data showing major population increases in East Coast striped bass; and the announcement of the birth of the first California condor chick conceived in captivity, an event that bodes well for the eventual restoration of the birds to the wild.

From National Wildlife, *February/March, 1989. Reprinted by permission.*

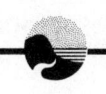

Earth Care Action

People Who Make a Difference

Citizens from all walks of life are taking extraordinary steps to safeguard wildlife and the environment.

Elanor Stopps: Seabirds' Best Friend

By Robert Boardman

Protection Island rises fortresslike from the waters off the north coast of Washington State. Two sand peninsulas, steep cliffs, and loose glacial soil make the windswept island an ideal nesting site for an amazing variety of seabirds that includes tufted puffins, pelagic cormorants, Pacific oyster-catchers, and the world's fourth largest colony of rhinoceros auklets. In all, nearly 75 percent of the seabirds found in Puget Sound nest on the island.

"There's hardly a square foot of land that isn't used by some bird during breeding season," says Elanor Stopps, a 69-year-old homemaker from the Olympic Peninsula community of Port Townsend, who led a 14-year effort to keep the island out of the hands of developers.

That effort culminated last summer, when the Protection Island National Wildlife Refuge was officially dedicated by the U.S. Fish and Wildlife Service. It was the first such refuge established during the years of the Reagan administration. "Nothing was more important [in the creation of the refuge] than the tireless devotion and dedication of one person—Elanor Stopps," observes former Rep. Don Bonker (D-

Elanor Stopps at Puget Sound's Protection Island National Wildlife Refuge, which she helped create. Stopps led a 14-year battle to keep the seabird nesting site out of developers' hands. (Paul Boyer)

Wash.), who had introduced the refuge bill into Congress.

Stopps began visiting Protection Island in the mid-1960s with her friend Zella Schultz, an ornithologist who was studying the birds there. "I'd help her band gulls for the Fish and Wildlife Service and we would talk about what we could do to help protect the island for the birds," says Stopps. Soon it was evident that the birds did indeed need help.

In 1974, after Schultz passed away, Stopps decided that it was up to her to do something about getting the island under public control. A developer had purchased the entire 400-acre land mass and was slicing it into 900 tiny lots under the guise of a retirement community, even though there was virtually no good water source on the island.

"The plan was simply unbelievable," says Stopps. "Can you imagine the impact of 900 septic systems alone? I wasn't an activist in one sense of the word. I was a housewife and a Girl Scout leader. But I had to learn."

Stopps soon realized that she had to build a base of support if she was ever going to interest the federal government in gaining control of the island. She wrote hundreds of letters and made thousands of phone calls to conservation groups around the country. She also created an "Adopt a Seabird" campaign that in time raised more than $50,000 for the effort to buy back the island.

Eventually, she enlisted the aid of Bonker, who brought the measure before Congress. "When the Reagan administration took power, all the professional lobbyists told me the refuge was dead," says Stopps. "But I just couldn't give up." She didn't.

In 1982, President Reagan signed the refuge into law. It took another six years, however, to secure enough funds to buy back the entire island. "Now I can relax a little," says Stopps, adding: "You know, there's nothing all that special about me. I just think I am an example of what ordinary people can accomplish if they truly believe in what they are doing."

Rolf Benirschke: Cans for Critters

By Michael Tennesen

When Rolf Benirschke was selected earlier this year as the new daytime host of the popular "Wheel of Fortune" television game show, few people in San Diego were surprised. After all, for ten years now, the affable, 34-year-old former professional football player has been one of the California seaside city's most popular—and accessible—local celebrities.

His popularity, however, is not pegged so much to his athletic achievements as it is to his ceaseless efforts to gain greater support for the San Diego Zoo's endangered species breeding and education programs. In recent years, Benirschke has devoted hundreds of hours to talking with schoolchildren and community groups about the plight of wildlife and getting those groups to help raise funds—literally millions of dollars—for the zoo's Center for the Reproduction of Endangered Species (CRES). And despite his recent entry into the world of show business, Benirschke is more active than ever in promoting conservation programs in Southern California.

"The zoo is blessed to have Rolf's help," says Betty Jo Williams, president of the San Diego Zoological Society.

Benirschke's interest in wildlife was fostered in part by his father, Kurt, a scientist who for 12 years was research director at the San Diego Zoo. A talented placekicker, the younger Benirschke turned down several scholarship offers from major football powers to attend the University of California at Davis, where he earned a degree in zoology.

Later, after joining the San Diego Chargers in the National Football League, he initiated a program called "Kicks for Critters," in which supporters agreed to donate funds to CRES for each field goal Benirschke kicked successfully. "My goal," says Benirschke, "was not only to raise money but also to raise awareness of the problems associated with endangered animals."

By the early 1980s, the placekicker knew he had made headway into achieving that awareness after a tough game in which he missed four field goal attempts. "The next day," says

Former pro football star Rolf Benirschke eyes a captive, endangered coatimundi. Benirschke heads up a successful San Diego school recycling effort that raises funds for endangered species research. (Michael Tennesen)

Benirschke, "the headline in the local newspaper read: 'There Were No Kicks for Critters.'"

Fortunately for wildlife and the Chargers, Benirschke did not have many days like that one. When he retired from football two years ago, he left as the Chargers' all-time leading scorer with a total of 766 points. And during the seven years of "Kicks for Critters," he raised more than $1.3 million for CRES.

Almost immediately upon retiring, Benirschke initiated a new program called "Cans for Critters," in which he began seeking the aid of San Diego schoolchildren. Now, every spring, students at more than 100 area schools collect aluminum cans for recycling, with the money raised from those collections going to zoo efforts. To get students interested, Benirschke makes appearances at schools.

Says zoo spokesperson Gabriella Green: "He makes children realize that they can do something that will help make a difference in this world."

Ila Loetscher: The Ridley's Best Friend

By Janice Arenofsky

People who live on Padre Island, off the Gulf Coast of Texas, call her "the turtle lady," and for good reason: For more than two decades, 83-year-old Ila Loetscher has spent almost all of her time trying to save the Atlantic ridley, the world's most endangered species of sea turtle. Though the animal's numbers have continued to decline, Loetscher has refused to give up. Without her efforts, the situation might be much worse.

The energetic senior citizen pioneered work on the ridleys that eventually stirred the U.S. government into action. "She doesn't kid people that she's a zoologist," says David Owens, a biology professor at Texas A&M University. "But by trial and error, she's become an expert."

Over the years, Loetscher has done everything from gathering turtle eggs from nests on the beach and instigating captive breeding efforts to rehabilitating hundreds of injured ridleys. These days, she spends much of her time lecturing about the turtles to schoolchildren and Padre Island tourists.

"Ila has spent many years doing great work," says Jack Woody, National Sea Turtle Coordinator for the U.S. Fish and Wildlife Service. Unfortunately, he notes, the ridley's breeding population has been declining 3 to 4 percent each year. The problem: The species has only one remaining nesting ground—Ran-

Ila Loetscher, who pioneered efforts on the Texas Gulf Coast to help save the world's most endangered sea turtle, clutches an injured, captive Atlantic ridley. (Evelyn Sizemore)

cho Nuevo, a mile-long strip of Mexican coast located 110 miles south of Texas. There, for years, poachers have been killing the turtles for their shells, meat, and hides, and then taking the animals' eggs.

Loetscher first heard about the situation at Rancho Nuevo in the mid-1960s, after she had moved to Padre Island from New Jersey. Encouraged by the notion that adult turtles might return to nest on the same beach where they had hatched, she decided to transplant ridley eggs from Mexico at her own expense to a more protected site on Padre Island.

"A whole group of us drove down to show the Mexican officials we were in earnest," she says. "We put about 2,000 of the ridley eggs in one big box and took them through customs at Brownsville, Texas. Then we buried them about eight miles from my home." Fifty to 70 days later, the eggs hatched and the young turtles quickly scurried to sea.

Ten years after this 1967 milestone event, the U.S. government joined the "Save the Ridleys" movement. "I don't think we would have been encouraged to do the Headstart program," says David Owens, "if Ila hadn't already done it on a modest scale."

Operated by the federal government with the cooperation of Mexican officials, Headstart now airlifts some 2,000 ridley eggs annually to Padre Island. There, the eggs are incubated, and the hatchlings are then nurtured for a year before they are released to sea. Loetscher is hopeful that some of those turtles eventually will return to Padre Island as nesting adults. And when they do, she says, she will be there waiting for them.

Meanwhile, last March [1988], in recognition of her efforts to try to save the critically endangered Atlantic ridley, the National Wildlife Federation honored Loetscher with a Special Conservation Achievement Award.

Art Aylesworth: On the Trail of Bluebirds

By R. Stephen Irwin

"With their feathers all ruffled out against the cold, they looked like big blue Christmas ornaments," says Art Aylesworth, recalling a March day 14 years ago when he spotted some mountain bluebirds perched in a thicket of fir trees near his western Montana home.

The incident left a lasting impression on Aylesworth, a Montana native who hadn't seen

In the 14 years since Art Aylesworth built his first bluebird nesting box in Montana, he has organized a broad network of volunteers that has put up thousands of the boxes to aid the birds. (Alan and Sandy Carey)

many bluebirds since his childhood. "They used to nest in our family's mailbox," he says.

That spring, the hard-working, independent insurance agent built five bluebird boxes, hoping to entice some of the creatures to nest near his house. He was fortunate. Two mountain bluebirds soon located one of the boxes and fledged all of their young in it. Aylesworth was hooked; the western and mountain bluebirds have been reaping the benefits of his enthusiasm ever since.

Members of the thrush family, the three species of bluebirds are found only in North America, where they are distinguished by their ranges and the males' coloration. The most well-known species, the eastern bluebird, breeds largely from the Rocky Mountains to the Atlantic Ocean. The western bluebird ranges in river bottoms from the Pacific Coast to the Rockies. The mountain bluebird is found in many mountainous areas of the West, from Alaska to Mexico.

All three species are cavity nesters, raising their young in abandoned woodpecker holes in trees, wooden fence posts, and utility poles. In recent decades, all three have suffered from a shortage of suitable nesting sites—a problem compounded by the introduction to this country of aggressive, competing species like the house sparrow and starling. In the West, the gradual replacement of wooden fence posts with metal ones has made matters even worse in many areas.

Aylesworth decided to do something about the situation. In the mid-1970s, he formed a group—Mountain Bluebird Trails—and began enticing friends and neighbors into helping him build nesting boxes. Soon, they were establishing "bluebird trails" in the region—strings of nesting boxes placed near each other in the proper habitat. He talked local timber company officials into donating the wood, and got high school shop classes to precut the lumber.

"Art has the ability to get people involved," says Montana businessman Gary Granley, who volunteered to help. "You can say 'no' to him and he won't take it personally. But then he'll come back at you with a different approach."

Eventually, a network of about 150 volunteers put up, and continue to tend to, more than 18,000 boxes throughout the state. "The western bluebird in particular could have been lost as a species in Montana had it not been for this effort," says Dennis Flath, nongame coordinator for the Montana Department of Fish, Wildlife, and Parks.

For the past decade, Aylesworth has also been traveling throughout the region, giving some 20 slide shows a year to various groups. As a result, other volunteers are now putting up nesting boxes in southern Alberta, Idaho, Wyoming, Utah, Nevada, and Washington.

For the 60-year-old Aylesworth, all of the hard work and travel clearly have paid off. Fourteen years ago, his first nesting boxes produced five young. Last year, Mountain Bluebird Trails helped fledge 8,195 of the bluebirds in western Montana alone.

Jan Garton: Keeping the Wetlands Wet

By Randall Winter

The first time Jan Garton visited the Cheyenne Bottoms Wildlife Area in central Kansas, she was hardly impressed. "I hated it," she says of a 1970 ornithology-course trip to see birds in the 19,000-acre wetland. "But later, as I learned to identify the many species at the marsh, I realized what a gold mine the place is."

Unfortunately, Garton also learned that the

WILDLIFE

Almost single-handedly, conservationist Jan Garton has helped ensure the vitality of one of the nation's most important waterbird staging areas: the Cheyenne Bottoms wetlands in Kansas. (Randall Winter)

gold mine was in danger of drying up, as its main water source was being depleted before it reached the marsh. Since then, the tenacious, 38-year-old Manhattan, Kansas, native has almost single-handedly changed the fortunes of this vital wetland habitat.

"In the past," says Stan Wood, former manager of the Cheyenne Bottoms Wildlife Area, "the Bottoms has had a lot of users—but not many supporters. Jan Garton brought the issue of saving the wetlands to the forefront, and kept it on people's minds." In doing so, she also helped ensure the survival of hundreds of thousands of migrating birds.

Historically, Cheyenne Bottoms may have provided a staging area—a place for migrators to rest and feed—for more shorebirds than any other site east of the Rockies. "It was and probably still is the most important waterbird area in the Central Flyway," says Marvin Schwilling, nongame project leader for the Kansas Department of Wildlife and Parks.

During spring censusing, authorities have found that about 45 percent of all North American shorebird species utilize the Bottoms, and more than 90 percent of the individuals of at least five species depend upon it as a staging area. Hundreds of thousands of ducks and geese also stop there. What's more, the Bottoms is considered crucial to the survival of the endangered least tern, and it has provided feeding grounds during part of the year for nearly one-third of the world's population of whooping cranes.

Because of its importance, Cheyenne Bottoms has been protected as a state wildlife area for decades. And though water was never in great supply there, it did arrive in sufficient quantity in most years—via the Arkansas River—to replenish the marshes for migrating birds. More recently, however, agricultural practices have reduced the Arkansas to a trickle by the time it reaches the Bottoms. In the early 1980s, the wetlands were receiving less than

10 percent of the water they were legally entitled to under past agreements. Enter Jan Garton.

In 1983, after hearing that the Central Flyway oasis was drying up, she took it upon herself to organize a conference of conservationists and state officials to try to solve the dilemma. Then she began lobbying legislators, who eventually appropriated $75,000 from the state treasury to fund a feasibility study for restoring water supplies to the marshland. It was the first time that Kansas had allocated general revenue monies for a natural resources conservation effort. In honor of her work, the Kansas Wildlife Federation, a National Wildlife Federation affiliate, named Garton its Conservationist of the Year in 1984.

Garton then organized a second conference to discuss the results of the feasibility study, and what might be done to implement them. "The Bottoms will never be fully protected until the state comes to terms with water usage and allocation," she says. "Unfortunately, no irrigator values wildlife on a par with crops."

Though Cheyenne Bottoms is by no means out of danger, enough water is now reaching the area to at least maintain its vitality.

Fred Slayden: Island Watchdog

By Mark Wexler

Bird watching has always been Fred Slayden's favorite hobby. So it seemed appropriate that the former Seattle, Washington, resident would begin spending his free time looking for feathered creatures in the bays and forests of St. Croix, soon after moving there in the late 1970s. A mechanical engineer by training, Slayden assumed that people on the 23-mile-long Caribbean island generally knew what species were found there. He was wrong.

"No one, not even authorities from the U.S. Virgin Islands Division of Fish and Wildlife, had an up-to-date checklist," says the 52-year-old Slayden. He decided to do something about it on his own. Before long, he had compiled detailed records of his sightings of 176 species of birds, including nearly 40 never before recorded on St. Croix. Then, with a notebook full of ideas about protecting wildlife habitat and his bird data in hand, he went looking for a local conservation group to join. There wasn't one.

"I contacted people at the Virgin Islands Conservation Society, National Wildlife Federation's affiliate based on St. Thomas, and asked them to help set up some meetings on St. Croix," says Slayden. "They said I could have all the meetings I want. 'As of this moment, you're the chapter president,' they told me."

That was in the spring of 1986. Since then, under his leadership, Slayden's chapter of the affiliate—called the St. Croix Environmental Association—has grown from a membership of 1 to 420. It also has become the island's leading conservation watchdog.

"Almost overnight, that group has emerged as an important political force on St. Croix, partly because its diverse membership cuts across racial and economic lines," says John Ogden, a biologist who serves on the island's Coastal Zone Management Commission. "It has filled a gap that's been missing here for a long time: citizen participation in decisions about island development."

"The people of St. Croix are proud of their natural resources, and for a long time, the local government has had the laws it needs to protect wildlife and habitat," adds Slayden. "But unfortunately, it has long lacked the data to support all of its decisions. We're trying now to help solve that problem."

Recently, Slayden presented some of his data to the Coastal Commission, which must approve all coastline development. At issue was a proposed housing project at a wetland area called Southgate Pond. A study by Slayden found that more than 50 percent of the bird species that breed on St. Croix depend on wetlands, and that Southgate Pond is one of the most important breeding areas in the Virgin Islands. The Commission voted unanimously against the project. Such a vote, says Slayden, "would have been unheard of a few years ago."

The conservationist made a lifelong dream come true not long ago, when he gave up a high-paying job at a local refinery to do biological fieldwork full-time for the Virgin Islands Division of Fish and Wildlife. "Imagine," he says, "getting paid to watch birds."

From National Wildlife, *June/July 1988 and August/September 1989. Reprinted by permission.*

A Citizen's Guide to Personal Action

ACTION ALERT

Watching the Watchdogs: How to Influence Federal Agencies

"If people will take the initiative, they can influence the decisions the federal government makes about our natural resources," says Lynn Greenwalt, vice president of the National Wildlife Federation's Resources Conservation Department and former director of the U.S. Fish and Wildlife Service.

Federal agencies such as the U.S. Fish and Wildlife Service, the Environmental Protection Agency, the U.S. Forest Service, and the Bureau of Land Management make many of the day-to-day decisions about how our air, land, water, and wildlife will be managed.

You can play an important role in this decision making! This section will show you how to use the system so you can keep an eye on federal agency activities and speak up if you see something that should be changed.

Making the Rules

Federal agencies have enormous power and responsibility. Congress may pass our laws, but the agencies establish the rules and regulations that put these laws into practice.

For example, Congress passed the Clean Water Act in 1972. Among other things, the law says "it is a national goal that the discharge of pollutants into navigable waters be eliminated," and directs states to set water quality standards.

But it is left up to the Environmental Protection Agency to draw up the detailed regulations that help the states meet these water quality goals.

Other federal agencies are responsible for carrying out the mandates of other environmental laws. For example, the Forest Service decides whether a particular stand of timber in a National Forest will be cut; the Fish and Wildlife Service determines when plants and animals should be listed as threatened or endangered.

"The Federal Agencies: Who's Who" on page 191 briefly describes some agencies and the work they do.

The Daily Newspaper for Federal Agencies

Suppose you want to know what plans the U.S. Fish and Wildlife Service has for Chinco-

The articles in this section are from National Wildlife Federation information booklets. Reprinted by permission.

teague National Wildlife Refuge. Or you want to know what controls the Environmental Protection Agency has placed on the manufacture of pesticides. Or you're interested in regulations pertaining to strip mining, marine animals, or forest roads.

One way to start is simply to pick up the telephone and call the appropriate federal agency. (The telephone book usually has U.S. government agency listings.)

"Another way to get information is to use the *Federal Register*," says Scott Feierabend, one of National Wildlife Federation's specialists on fisheries and wildlife regulations. "By using this publication, it's possible to follow the rule-making process of all 78 federal agencies."

The *Federal Register* is like a daily newspaper on federal agency actions. It contains official announcements of the proposed, interim, and final regulations made by each agency, and also lists executive orders, presidential proclamations, and other documents of public interest. In fact, most federal regulations cannot take effect unless they are published in the *Federal Register* first.

It is published Monday through Friday and can be found in most major public libraries.

The best time to make your voice count is when a regulation is still in the proposed stage. (You can comment on a rule after it has become final, but at that point, an agency is no longer legally required to consider your concerns.)

"Public comment does influence an agency's decision," says Feierabend. One example, Feierabend recalls, was a petition filed by concerned citizens that forced the National Oceanic and Atmospheric Administration to change the way in which shrimpers could legally set their nets. The old nets were trapping endangered sea turtles; the new nets allowed the turtles to swim free.

Sharon Newsome, Director of Legislative Affairs for the National Wildlife Federation, recalls how the federation acted on another public comment notice published in the *Federal Register*.

"We urged our members to write letters in opposition to a particular regulation of the newly enacted Coastal Barrier Resources Act," Newsome said. "The mail received by the Flood Insurance Administration was 20 to 1 against this proposed regulation. As a result, officials of the agency changed their minds and the final regulation was never issued."

How to Use the *Federal Register*

The *Federal Register* is really quite easy to use if you begin by looking at the table of contents and index in each issue.

For example, the April 8, 1990, table of contents might show that the Environmental Protection Agency is proposing a new regulation on toxic chemical "ABC." Page 11947 (page numbers are cumulative) of the day's *Register* explains that the agency has extended the time for public comment on the proposed regulation.

First, the regulation is described briefly. You learn that it originally was published in full on page 3738 of the January 29, 1990 *Federal Register*. An address is provided so you can send written comments, and you see that public hearings on the topic will be held in Washington, D.C., beginning July 14, 1990.

When to Use the *Federal Register*

Of course, most people don't have time to look through the *Federal Register* every day. But there are two times during the year when you should be sure to look at the publication.

In April and October, a unified agenda is published. This agenda outlines the long-term plans of all the federal agencies and describes

ACTION ALERT

what types of issues they will be looking at and what the time schedule will be. Knowing what's in the works will help you know when to check future issues of the *Federal Register*.

Other Ways to Find Information

Articles in your local newspaper also give you valuable clues about when to write to federal agencies.

One way to obtain more information on an issue that is reported in a local newspaper would be to look at issues of the *Federal Register* for about two weeks before the local article appeared to find the full text of the proposed regulation.

Also, you could call the district offices of the specific agency involved. (Check your telephone book for the phone number.)

The Federal Agencies: Who's Who

When you're concerned about an issue and you want to express your opinion, you should know whom to contact to get the best results. Here's a brief rundown of the agencies and what they do.

Department of the Interior

The Department of the Interior, which is responsible for our country's natural resources, is made up of approximately 30 major bureaus and offices. Among the offices that are important to conservationists are the U.S. Fish and Wildlife Service, the National Park Service, the Bureau of Land Management, the Office of Surface Mining, and the Bureau of Reclamation.

The *U.S. Fish and Wildlife Service* is the principal agency through which the federal government helps to conserve our nation's wildlife—from Kirtland's warblers to black-footed ferrets. In addition to managing the National Wildlife Refuges, the Fish and Wildlife Service oversees the National Fish Hatcheries. They also are responsible for the enforcement of several laws, including the Endangered Species Act and the Migratory Bird Treaty Act.

The *National Park Service* manages our National Parks, from Arizona's Grand Canyon National Park to Acadia National Park along the coast of Maine. The White House, the Jefferson Memorial, and various other national monuments also are administered by this agency. In addition, the Park Service coordinates the Wild and Scenic Rivers System and the National Trail System.

The *Bureau of Land Management* oversees approximately 300 million acres of public lands, mostly in ten western states and Alaska. These public lands are managed for a variety of uses, including outdoor recreation, fish and wildlife habitat, livestock grazing, timber, and watershed protection.

The *Office of Surface Mining* administers the Surface Mining Control and Reclamation Act of 1977. This act set up the environmental regulations that allow the land to be strip-mined without permanent damage to the soil and water. The office also helps coal-mining states comply with surface mining laws and collects money from coal companies to fund reclamation projects.

The *Bureau of Reclamation,* established in 1902, handles western water projects, which provide irrigation water for farms and drinking water for cities and towns, plus flood control and hydroelectric power. The bureau's projects have included Grand Coulee Dam and Hoover Dam.

Environmental Protection Agency

The Environmental Protection Agency (EPA), which was established in 1970, enforces many of our major environmental laws including the Clean Air Act, Superfund, the Safe Drinking Water Act, the Clean Water Act, and the Federal Insecticide, Fungicide, and Rodenticide Act. Divided into ten regional offices, the EPA helps states comply with environmental laws and develop cases against polluters. The agency also conducts research on the causes, effects, and control of environmental problems.

Department of Agriculture

The U.S. Forest Service and the Soil Conservation Service are two agencies within the Agriculture Department that administer our environmental laws.

The *U.S. Forest Service* manages our National Forests. They range from Alaska's 17-million-acre Tongass National Forest to Puerto Rico's Caribbean National Forest, which is one of the few tropical rainforests found in the United

(continued)

How to Write to Federal Agencies

Although you don't have to be an expert to comment on an agency's proposed regulation, there are steps you can take to make your comments as effective as possible.

First, type your comments and be sure to include your name and address. Also, specify exactly which regulation you are commenting on. Sometimes the agency will give additional directions in the *Federal Register* about specific information that is sought and how to label and mail your comments.

The Federal Agencies: Who's Who—Continued

States. Our National Grasslands also are under the jurisdiction of the Forest Service. In addition, the agency oversees road construction, timber harvesting, and mining in the National Forests.

The *Soil Conservation Service* was established in 1935 to promote soil and water conservation. The agency gives technical and financial assistance to farmers, ranchers, and state and local governments to reduce soil erosion, prevent floods, and conserve water. The Soil Conservation Service also conducts research to develop new strains of grasses, legumes, and trees for a wide range of uses.

Department of Commerce

Two environmental agencies within the Commerce Department are the National Oceanic and Atmospheric Administration, a major federal environmental management agency, and one of its divisions, the National Marine Fisheries Service.

The *National Oceanic and Atmospheric Administration (NOAA)* keeps an eye on our environment from the highest clouds to the deepest oceans. The duties of NOAA personnel include predicting the weather, making nautical charts and coastal maps, helping states protect their coasts, using deep-sea vessels to explore the ocean floor, and monitoring marine pollution.

The *National Marine Fisheries Service*, a division of NOAA, protects whales, porpoises, sea turtles, seals, and sea lions. Among other things, the agencies' biologists study the populations, life cycles, and breeding habits of these marine mammals. The National Marine Fisheries Service administers the Marine Mammal Protection Act and the Fishery Conservation and Management Act.

Department of Defense

The U.S. Army Corps of Engineers is a division of the Department of the Army, within the Department of Defense. Besides planning and building dams, reservoirs, levees, harbors, and locks, the corps has main responsibility for Section 404 of the Clean Water Act, which is the nation's most important wetlands protection law.

Department of Transportation

Two agencies within the Department of Transportation that handle environmental issues are the Coast Guard and the Federal Highway Administration.

The *Coast Guard*, the federal government's primary maritime law enforcement agency, administers the oil spill liability section of the Clean Water Act. The agency also inspects ships that carry oil and issues certificates of responsibility to owners and operators of vessels that are liable for oil spills.

The *Federal Highway Administration*, in addition to constructing and maintaining our national highways, also designs and builds the roads in our National Parks. In addition, this agency is responsible for the Highway Beautification program and regulates the movement of dangerous cargo—including hazardous chemical wastes—on our highways.

Department of Energy

The Department of Energy, which includes the Office of Conservation and Renewable Energy and the Federal Energy Regulatory Commission, was established in 1977 by President Jimmy Carter to bring together in one central Cabinet-level agency all of the federal government's energy responsibilities.

The *Office of Conservation and Renewable Energy* directs the Department of Energy's energy conservation programs. It also conducts research in solar and renewable energy resources.

The *Federal Energy Regulatory Commission* is an independent regulatory agency that is within, but separate from, the Department of Energy. Established in 1977 to replace the Federal Power Commission, the agency licenses the construction and operation of nonfederal dams and other hydroelectric projects.

ACTION ALERT

When commenting on a regulation, you do not have to use technical language. For example, if you are opposing the use of off-road vehicles at Assateague National Seashore, all you have to do is write a letter that says you don't want ORVs on the beach. If you can, explain how and why the proposed rule will affect you.

Remember, it can be just as important—sometimes more important—to write to federal agencies as it is to write to your Representative or Senator. Because federal agencies do not receive as much mail as Congress, individual letters have much greater impact. Also, federal agencies are required by law to solicit public reaction, and they do weigh it carefully.

Environmental Impact Statements: Your Roadmap for Federal Projects

There is another way you can get involved in federal decision making—at a local level.

Suppose a timber company wants to cut trees in the National Forest near your home. The responsible federal agency—in this case the Forest Service—will have to make a decision: Should the company be allowed to cut the trees, and if so, how many and in what manner?

At this point, the local office of the Forest Service may prepare an environmental impact statement, which, among other things, assesses the ecological values of the forest and the effect that timber harvesting would have on those values.

The environmental impact statement is written by biologists, economists, and other federal government experts. But the public has several opportunities to participate in the process.

While the Forest Service is preparing the statement, they may decide to hold meetings. You can find out about these meetings by calling the local office of the Forest Service and looking for announcements in the local newspaper.

You don't have to be an expert to speak at a meeting. Your ideas and opinions are what really matter. If you oppose the timber harvest as proposed, tell the Forest Service why. Will timber cutting harm an endangered bird, ruin a campground, or hurt hunting? Be specific in your comments, and if possible, suggest alternatives.

Another good way to influence the process is by informal meetings with those who are preparing the document. You may find out that they are unaware of information you may have or have failed to consider important questions. Also, it is a good idea to make sure that the local media know about your concerns.

You Can Make a Difference

"People shouldn't be afraid to write a letter or speak to agency officials," says Lynn Greenwalt. "These actions often make a real difference when an agency makes its final ruling."

"If you get involved with some vigor and perseverance," Greenwalt concludes, "there is an excellent chance you will influence the actions of the federal agencies."

Remember, action taken by the Interior Department, Environmental Protection Agency, and other federal agencies does affect all of us. It's up to us to make sure these agencies do their jobs well.

Use the News to Protect Your Environment

Suppose you wake up one morning, turn on your water faucet, and the water is mysteriously murky. Or you look out your window some afternoon and discover a bulldozer plowing through your child's favorite woods to make way for a new highway.

What do you do? Can one person really do anything to stop a highway? Or protect their drinking water?

The answer is yes. Just ask Lois Gibbs, who single-handedly brought national attention to the dangerous toxic wastes abandoned at Love Canal. Or Annie Snyder, who is credited with stopping the construction of numerous ill-planned shopping centers and trailer parks.

One person, with little or no experience as an activist, *can* start a conservation fight and win. But to be successful, you must know how to publicize an issue.

Here, we'll show you how to use both print and broadcast media to call attention to environmental problems. We'll tell you whom to contact at your local newspapers, television stations, and radio stations, and how you can obtain the very best results.

We'll help you challenge that construction project or get to the bottom of the murky tapwater—and win.

Newspapers

Your local newspaper is the ideal place to begin. There are several ways your story can find its way into print.

Letters to the editor. "One of the easiest ways to get your problem or story into a newspaper is to write a letter to the editor," advises the National Wildlife Federation's media specialist.

The Letters to the Editor column is one of the most influential and widely read sections of any newspaper.

A carefully written, one-page letter will alert editors, journalists, and the public to your problem.

For example, Tom Austin, a grassroots activist in Indiana, has launched several conservation fights against air pollution by using letters to the editor.

"I frequently use letters to the editor to alert others to environmental issues I feel strongly about," says Austin. "My letters have motivated other people to write letters to the editor in support of my cause. Then the editors realize other people are concerned as well, and they begin to cover my story."

News stories. A story in the newspaper will not only call attention to your problem, but will tell the rest of the community that your story has credibility.

If you feel your issue deserves local news coverage, don't be afraid to call a reporter.

"People are our stock and trade," says Paul G. Hayes, science reporter for the *Milwaukee Journal*. "Half the time we [reporters] don't know what's going on out there unless people call and tell us," he says.

When talking with a reporter, Hayes advises:

- Your issue should be timely; that is, it should be happening now.
- Your issue should be important to others in the community.
- You should be armed with the facts. Document your story whenever possible with tangible proof such as photographs and samples.

Before you call a newspaper reporter, it's a good idea to know the newspaper's deadline and the name of the reporter you need to talk to. To find out the deadline, simply call the newspaper and ask. (It's best not to call a reporter right before the paper's deadline.)

To pinpoint the right reporter, look through your local newspaper for articles on local environmental or energy issues. When you find an appropriate article, look for the byline. That's the person you talk to.

They may go by many different titles, including environmental reporter, science reporter, outdoor reporter, or energy reporter. If your paper does not have reporters that specialize in different areas, we suggest you contact a general reporter.

When you reach a reporter, say who you are, where you live, and what's wrong. Remember to tell your story with facts and to tell how it affects others. If it is "newsworthy," the reporter will take it from there. And before you hang up, make sure the reporter knows how to reach you in case there is anything else you can do to help.

Sometimes a reporter will want to meet with you. If possible, put together a packet of information, including a fact sheet of dates, places, and events, and include photographs if possible.

Linda Niccum, a 66-year-old grandmother, started a campaign to clean up a toxic dump site in her home town of Greenup, Illinois. She is convinced that she got things done by using newspaper reporters. "Chemical companies don't like newspaper stories," she says.

Feature stories. Don't be discouraged if a news reporter doesn't think he or she can use your story. In that case, your next step should be to call a feature reporter.

Although you should approach a feature reporter the same way you would a news reporter, you should realize that the resulting story will be different from a "hard" news story.

In the case of the highway construction project, for example, a news story may focus on figures, events, and conflicting opinions. On the other hand, a feature story may explore the different plants and animals found in the woods and what happens to the life when bulldozers invade their habitat.

So, when talking with a feature reporter, emphasize the different aspect of your environmental issue. The result may be a lengthy, in-depth look at your problem—something a news story often doesn't do.

Editorials. Editorials become extremely important in a conservation fight because they are read by community leaders and public officials who are trying to keep up-to-date on issues in the community limelight. These people, in turn, can lend support that could be vital to your cause.

We suggest two ways of getting published in the editorial section: Write an editorial yourself, which is called an op-ed (opposite the editorial page) piece, or persuade the editor of the editorial section to write one for you.

"Op-ed pieces should be clearly and concisely written, typed, double spaced, and be no longer than two pages," says our media specialist.

Your article has a better chance of being published if it is written by a public official or other well-known person who supports your stand on the issue.

After writing your op-ed piece, submit it with a brief cover letter that includes some background information about your subject and your involvement in the issue. Address it to the editor of the op-ed page. (Again, you can get his or her name by calling the newspaper.)

Another approach to the editorial section is to call a newspaper and ask to speak with someone on the editorial page staff. Tell him or her your story. If that person is receptive and seems to support your position, offer to send all the background material you have on that particular issue.

Sometimes the newspaper will then run an editorial on the editorial page voicing its position on the subject.

Radio and Television

When starting your conservation fight, think of radio and television as just one more way to spread your environmental message.

"You may be a little hesitant when you first contact the media, but it'll pass," says Vic Barry, an individual who on several occasions has contacted television and radio stations for his fight against acid rain.

"Just pick up the phone and tell your problem to the first person who will listen," he advises.

There are a number of radio and television program formats you can use to generate public interest in your environmental problem, including newscasts, talk shows, public opinion messages, and editorials.

General tips for news coverage. Besides being the quickest way to draw attention to your environmental issue, news coverage will add legitimacy to your conservation fight.

"When dealing with the broadcast media, it is important that a person's story be newsworthy," says our media specialist.

For example, the best way to obtain coverage from a news department at a radio or television station is to emphasize:

● Something has happened today (your water is murky or it smells funny).

● You have some facts (you've had your well tested and your water is contaminated).

● This story is of interest to others (if your water is contaminated, it's usually not an isolated problem).

Radio news. The average radio station carries a five-minute newscast on the hour, with a 60-second update every half hour. Because so little time is devoted to news, most radio news is "hard" news. That means, it is happening now.

"Radio news directors will jump at the chance to cover a story that is newsworthy and hasn't hit the newspapers or television," says Susan Stolov, former news reporter for WISN, Milwaukee, Wisconsin. "In that case, a person is almost assured some sort of coverage."

When you call your local radio station, ask for the news director. Once connected, introduce yourself and tell him or her your story. Be prepared to answer questions over the phone. There is a possibility that you will be asked for permission to tape your interview so it can be used at a later time.

Television news. At television stations, you'll want to tell your story to the assignment editor. (They are the people who assign the reporters to cover a story.) Call your local television stations and ask to speak with the assignment desk.

When explaining your story, make it short.

Tell what will happen, where it will happen, when it will happen, and who will be affected.

If you can't get through to the assignment desk, we suggest you ask to speak with the reporter who covers environmental issues. Sell your story to that reporter and then let them sell your topic to the assignment editor.

Since a station's deadline is about one hour before its newscast, try not to call around this time. The best time to call is early in the morning.

Editorials. No matter what environmental issue you're fighting—toxic waste dumps or the destruction of forest land—editorials are an excellent way to introduce your issue to the public.

Call local radio and television stations and ask for the person in charge of editorials. At a radio station, you may end up talking with someone in the news department, because radio stations usually have small staffs and don't have a separate editorial department. But at a television station, ask to speak with someone in the public affairs department.

If a station decides to run an editorial on your issue, it will support or oppose your issue. If it supports your issue—great. But if it doesn't, you can request time to present a rebuttal.

This is the "equal time" rule, and it applies to both television and radio. It does not, however, apply to newspapers. If a newspaper comes out with an editorial against your position, you cannot demand equal space.

As soon as possible after the opposing editorial is aired, contact the person at the station in charge of editorials. Explain that you want rebuttal time. This is when you—or if you prefer, another person—tell your side of the story.

Another approach is to write a letter to a radio or television station about your concerns. Most radio and television stations encourage their listeners and viewers to respond to community issues; the end result could be a station's spokesperson reading your letter over the air.

Public opinion messages. Public opinions or free speech messages in radio and television are like op-ed pieces in newspapers—they present your side of the story. And, like newspaper editorials, the "movers and shakers" of the community pay attention to them.

The best way to find out if you can tape a free speech message and have it broadcast is to call a radio or television station and ask. At a television station, the public affairs department will handle these inquiries. At a radio station, it is best to let the receptionist connect you with the person who is in charge of public opinion messages.

Once you have reached the person who handles these messages, briefly explain your position and why it is important to the community. Remember, this message is not in response to an editorial, so the station is not obligated to run your statement. It is your job, once again, to convince the person on the phone to support your cause.

In Jackson, Mississippi, Tammy Reed has taped free speech messages, convinced station managements to support her cause through editorials, and written letters to radio and television stations—all in hopes of generating public support to restrict needless development on the barrier islands off the Mississippi mainland.

"If I call a station and talk to the person in charge of editorials, I find it's easier to get my cause on the air," she says. "If management doesn't want to take a position on what is happening with the islands, I write them letters or ask if I can tape an opinion message," Reed adds.

Talk shows. Don't overlook talk shows when you're trying to publicize a local environmental problem. We recommend talk shows, because issues are thoroughly discussed and explained—more than in any other broadcast format.

To find local talk shows, either check local newspapers for a schedule of nearby radio and television talk shows or call area stations and ask if they broadcast any talk shows.

Alternatively, you can check the *Broadcasting Yearbook* in your local library, which should contain a list of local radio and television talk shows.

Once you have found a station with a talk show, call that station and ask to speak with the producer of that show. Try to convince him or her to do a show on the subject about which you're concerned.

Suggest yourself and others as possible guests. Since talk shows like controversial subject matter, it also is a good idea to suggest people who oppose your position.

Follow that call with a letter. In the letter, restate your issue and provide a list of possible guests and their credentials. This is always beneficial because it gives the producer something tangible to refer to when he or she is deciding whether or not to do a show on your topic.

If you have secured a spot as a guest on a talk show—in the case of television—ask if you can bring visual aids such as slides, pictures, or a sample of your murky water. And for both radio and television, be prepared with facts, figures, and examples. One good way to combat any feeling of nervousness is by being prepared and rehearsing what you are going to say before you have to say it.

Sometimes talk shows open their phone lines. If you get on a call-in talk show, it's a good idea to listen to the show ahead of time to see what types of questions people may ask you.

We also suggest being a caller on a call-in talk show. If a talk show is discussing a subject related to your environmental issue, call in and participate. Because a lot of people listen to call-in talk shows, it's an excellent way to make people aware of your issue.

And remember—the next time an environmental problem rears its ugly head in your neighborhood, don't just sit there. There *is* something you can do. Get the media involved and you just might win that fight!

Making Congress Work for Our Environment

If you think you can't influence what goes on in Washington, consider this. Our country's rivers, streams, and lakes will be cleaner because people like you convinced Congress to pass the Clean Water Act over the President's opposition.

Every year, citizens concerned about their environment persuade Congress to pass needed laws to protect our air, land, and water. Your legislators in Washington will listen to you because your views provide first-hand information on how an issue affects the people "back home."

Here we'll show you how to contact Congress effectively. Included are guidelines on how to write a letter to your legislator that will make it stand out from the thousands of other letters received every day. We tell you how to

ACTION ALERT

send a telegram, how to use your phone, and how to set up a meeting with your representative. "Understanding the Congressional Process" below also gives information to help you better understand how laws are passed.

Writing Letters

It only takes a few minutes to write a letter, but those few minutes can make a big difference. When members of the House or Senate receive enough letters on a particular issue, it *does* influence their vote. After all, they want to be reelected, and your vote counts.

Here are some suggestions that will give your letter the greatest possible impact.

Use your own words and your own stationery. A handwritten letter is fine, as long as it's legible.

Be concise. A one-page letter is long

Understanding the Congressional Process

Understanding how legislation becomes law can be a challenge, but it is useful. The better you comprehend the whole process, the more effective you can be at persuading your members of Congress.

Congress refers collectively to both the House of Representatives and the Senate. Each state is represented by two senators and by a number of representatives proportional to that state's population. House members are elected every two years, while senators are elected every six years. One third of the Senate comes up for election every two years.

The legislative process begins when a bill is first presented, or introduced, to the House or Senate by a sponsoring representative or senator. First, the bill is given a number. Bills are prefixed with "H.R." when introduced in the House, and with "S." when introduced in the Senate. These initials are followed by the assigned number; this number reflects the bill's order of introduction in each two-year congressional session.

Committees

Next, the bill is referred to a committee and a subcommittee. Committees are divisions within both the House and the Senate that deal with specific kinds of legislation. For example, the Committee on Energy and Natural Resources considers, or has jurisdiction over, legislation on energy conservation, energy regulation, and public lands. Each committee is further broken down into subcommittees. For example, the Senate Committee on the Environment and Public Works is divided into six different subcommittees, including the Subcommittee on Environmental Pollution and the Subcommittee on Water Resources.

Often more than one committee will have jurisdiction over a bill, and the bill may then be considered by several committees. For example, if a representative drafts a bill dealing with marine pollution, the bill could be referred to both the House Committee on Merchant Marine and Fisheries and the House Committee on Interior and Insular Affairs.

Because more than 10,000 bills are introduced in each Congress, if a bill does not receive substantial support, no action will be taken. When this happens, the bill is said to have died in committee. But if the legislation gathers enough advocates in Congress and is sponsored by the committee or subcommittee chairman, subcommittee hearings will be scheduled to review the bill's merits and flaws. Members show their support for legislation by putting their names on the bill as cosponsors.

Hearings

When hearings begin, the subcommittee usually invites testimony from interested witnesses, although some people testify at their own request. Witnesses may include legislators who support the bill; officials of government agencies, such as the Department of the Interior or the Environmental Protection Agency; interested organizations, such as the National Wildlife Federation; experts, such as college professors; and individual citizens.

To find out if legislation is scheduled for action, you can contact the legislation status office. (Telephone numbers are listed on page 203.)

Markup

After gathering suggestions and information from these hearings, legislators hold subcommittee meetings to consider "markup" of the bill. During markup, changes and additions are made to the bill's original text through amendments, which may be proposed by any

(continued)

enough; it is more likely to be read than a two- or three-page letter.

Try to write more than just one or two sentences, however. Instead of writing, "Vote for H.R. 544" and nothing more, explain how the issue will directly affect your life or the people and resources in your area. Personal touches will often influence congressional representatives.

Identify your subject clearly. If possible, try to refer to legislation either by its bill number or its popular name, such as the Endangered Species Act.

Discuss only one issue in each letter. For example, don't mention acid rain and hazardous wastes in the same correspondence. This ensures that your letter will be seen by the right staff member.

Understanding the Congressional Process— *Continued*

member of the subcommittee. Next the bill goes to the full committee, where further revisions may occur during-markup. (Letters to a subcommittee or full committee are most important and influential just before these markups.) The full committee may approve, or "report," the bill and recommend passage, or it may not take any action at all, thus killing the bill.

Floor Action

In the House, if the full committee approves the legislation, most of the bills will then go before the Rules Committee. This powerful committee decides how much time will be allowed for debate and also determines whether members of the House will be able to offer amendments when the bill reaches the floor of the House. The Speaker of the House decides when the bill will be scheduled for action.

In the Senate, there is no Rules Committee. Instead, the Senate allows unlimited debate, unless there is unanimous agreement on how much debate time should be allotted to a particular bill. Occasionally, they will even decide the number and type of amendments to be voted on.

The Senate Majority Leader decides when to schedule Senate action. Unlike the House, however, one senator can prevent action of a bill by placing a "hold" on the bill. Until the senator's concerns with the legislation are addressed, the Majority Leader is unlikely to ask for a Senate vote. If the bill is brought to the floor with a "hold" on it, the senator can "filibuster" or talk for as long as possible on the bill, thus preventing any votes. Because this delaying tactic can tie the Senate in knots and prevent action on other bills, the Majority Leader avoids scheduling bills with "holds."

Once a bill reaches the floor of either chamber, members either will accept committee recommendations or offer amendments. When all amendments are accepted or rejected, a vote for final passage is taken.

Conference Committee

When a bill on the same subject passes both the House and the Senate, there's often a difference between the two versions. When this happens, the chairmen of the committees that originally considered the two bills put together a conference committee made up of a portion of the members from the original House and Senate committees. If the conference committee can't settle their differences, the bill dies. If they iron out the bill's areas of disagreement, the result is called a conference bill.

The new legislation is then sent back to both houses for a final vote and, if passed, is sent to the President. The President will either sign the bill into law or veto it.

If the bill is vetoed, Congress has one last chance to put the bill on the books— legislators can override a presidential veto by a two-thirds majority vote in both the House and the Senate.

Simply put, that's how a bill becomes law.

Appropriations

This process is called authorization. But there's one final step after authorization. If the legislation requires federal funding to put its programs into effect (as with toxic waste cleanups), then a separate appropriations bill must be passed to supply the necessary funds to implement the law. If no money is approved by Congress, or if funding is severely restricted, authorization bills can't do their job. Therefore, appropriation bills, passed every year to set the funding for federal programs, play a crucial role in the process of carrying out laws.

ACTION ALERT

Ask the legislator to do something specific. For instance, you could ask that he or she vote for a particular bill, request hearings, or cosponsor a bill.

Ask for a reply. With a response in hand, you will better know how to pursue the matter.

Try not to use form letters. They have one-tenth the impact of any letter you write yourself. Photocopies or carbons also make little impression. The same goes for postcards—they're considered a sign of organized "pressure mail" campaigns by most congressional offices.

You needn't mention anywhere in your letter that you're a member of a conservation organization. Legislators know where most major conservation groups stand on environmental issues. They want to know where *you* stand.

The National Wildlife Federation appreciates receiving copies of letters that you send to Capitol Hill. But you should avoid putting "cc: National Wildlife Federation" at the bottom of your letter.

Don't be unnecessarily critical, and never threaten or insult. Imagine how you would react if you received such a letter in the mail. Negative correspondence often does more harm than good, and can cause legislators to swing their support in the opposite direction.

Be sure to include your return address on the letter. Also, make a copy of your letter. It may come in handy later when you want to write again.

Sample Letter
Here is an example of an effective letter.

The Honorable Nancy Smith
United States Senate
Washington, DC 20510

Dear Senator Smith:

Please support S. 252, so we will stop polluting our groundwater before it's too late.

Within our small community, we're learning about groundwater pollution, but what we're learning is that no one knows enough. Our town's water supply was tested recently. Officials said they found nitrates in the water. The problem is they couldn't tell us exactly where these nitrates came from or what they will do to us. So we're left not knowing how afraid we should be—which is a pretty scary thing!

My husband and I have one small child, and we hope to have another. We want to raise them in a healthy world.

Please join in as a cosponsor of this groundwater research and control bill. Also, please keep me informed of your involvement with this legislation.

Sincerely,
Jane Brown
Address

Follow-up Letters
When your representative's reply to your letter arrives, it's time to take action again. Follow-up letters can be very important.

If your legislator agrees with your stand and plans to take action, write and thank him or her. Many letters, however, will not contain this response. Some of them will be noncommittal; they may fail to respond solidly to any of the points you make in your original letter.

If that is the case, you should write again and politely point out that your questions from your first letter were not answered, and you are still concerned and would like a response. Second letters will almost always get results—your legislator will know you're very concerned.

Also, send your thanks when a legislator's vote or other action supports your cause. Elected officials receive stacks of critical mail; a little praise now and then for a job well done is always welcome.

Using Your Telephone

Telephone calls are especially effective as a last-minute method of relaying your message to Washington. The cost is minimal (usually only a few dollars for a short call to D.C.), but it impresses a legislator when you're concerned enough to take time out from your busy day and money from your pocket to communicate how you feel.

Here are a few suggestions for telephoning your legislators in Washington.

Give your name, and ask first to speak with your legislator. State what issue you're calling about, such as "H.R. 2959, the pesticide reform bill Representative Jones will be voting on tomorrow." If your legislator isn't available, ask to speak with a staff member who specializes in the specific bill or environmental issues you're interested in.

Be brief, organized, and polite. State what you want the member to do—vote for or against a bill or specific amendment.

Note the name and the title of the person you speak with, as well as the date and the time of your call. If you ever wish to speak to your legislator or their staff in the future, such contacts can be invaluable.

If you wish, ask for any relevant materials to be sent to you, such as a copy of the bill.

You also can call your representative's or senator's local office. Representatives have at least one office in their congressional district; senators have one or two offices located in major cities in their state. Explain your views on the issue to local staff members and ask that they relay it to Washington. Contacting a local office, however, means that the message must go through several people before it reaches the member. So you should call three or four days before the vote.

Telegrams

Telegrams tend to be shorter, faster, and more urgent than letters. They offer another way to contact your legislators in the last few days before a crucial vote. Unlike telephone messages that may get lost in the shuffle, a telegram puts your opinion in writing. Although they tend to be more expensive than letters, some economical telegrams are available through Western Union.

Public Opinion Messages are one of the least expensive ways to contact officials in less than 24 hours. By simply phoning in your message to Western Union, you can send your views to Capitol Hill. Delivery cost is $8.95 for 20 words, and $2.50 more for each additional 20 words. There is no charge for name and address.

Twenty words can pack a powerful punch. Suppose you sent a message that read: "Support H.R. 3624 to control acid rain. Reduce sulfur emissions by 12 billion tons to save our lakes." That would certainly get someone's attention!

Mailgram messages are another fairly inexpensive way to contact your legislator. Your 50-word message (including the sender's name and receiver's names, addresses, and ZIP codes) will be sent to the nearest local post office and delivered the next day. Mailgrams cost $13.95 for 50 words, and $4.95 for each additional 50 words.

When it's important that your message arrive as quickly as possible, you should consider the standard telegram. Unfortunately, it's expensive. To send a telegram, just phone Western Union, 24 hours a day, seven days a week. Your telegram, which usually arrives within six hours, costs $25.90 for the first ten words and 65¢ for each word after that.

ACTION ALERT

If you need additional information, you can call your local Western Union office or call 1-800-325-6000.

Personal Meetings

One extremely effective method of communication is to meet with your elected officials. You can talk to your legislators when they return to their home district or, if you plan to be in Washington, go see them there.

To get an appointment, simply call the local or Washington, D.C., office and tell the staff member you'd like to meet with your legislator. If you don't know the location or telephone number of these offices, check with the reference section of your local library.

If approaching your elected officials by yourself is too scary, look for friends or other people who share your interest. Then approach your members of Congress together. The issue you're concerned about may affect a variety of groups, such as health organizations or unions. The wider the spectrum of people supporting an issue, the more weight your position will carry.

The House and Senate take periodic recesses and many lawmakers head home. One recess to count on is in August, when Congress takes a break from the hot and humid climate of Washington. Recesses are good times to catch up with your legislators and discuss the environmental issues they will be facing in Congress.

If you're not successful in securing a personal meeting, keep an eye out for notices of town meetings or public forums. Many legisla-

---

Useful Information

Addresses

The following basic addresses will help you direct your letters to the right location.

The Honorable ——
United States Senate
Washington, DC 20510

Dear Senator ——

The Honorable ——
U.S. House of Representatives
Washington, DC 20515

Dear Representative ——

Telephone Numbers

General Information
(202) 224-3121

This is the Capitol switchboard. Besides providing general information, they can connect you directly to your legislator and furnish needed phone numbers.

Legislative Status
(202) 225-1772

If you call this number, you can obtain the current status of legislation pending in the House and the Senate. They also can provide the bill number if you have specific information about a bill, such as its name and sponsor.

Cloakroom Tapes
Senate Republican Cloakroom:
  (202) 224-6391
Senate Democratic Cloakroom:
  (202) 224-4691
House Republican Cloakroom:
  (202) 225-7350
House Democratic Cloakroom:
  (202) 225-7330

These tape recordings cover current floor action and scheduling information.

tors may hold such public meetings during a recess, and you can at least ask a question on your environmental concern.

How to Get the Most out of Your Meeting

Make an appointment. To get an appointment, simply call your lawmaker's office and tell the staff member you would like a meeting with the legislator. Check with your local library for the number or call (202) 224-3121 (the U.S. Capitol switchboard) to speak with Washington, D.C., offices.

Be on time for your meeting. Legislators are very busy people. Chances are, you will not get that much time for your meeting, so you don't want to waste it.

Start with a compliment. Your meeting will go more smoothly if your legislator sees from the beginning that you are friendly. Start your meeting by expressing your thanks for a particular piece of legislation that has been sponsored in the past or for a good stand your representative has taken on an issue.

You don't have to be an expert. You will probably know much more about an issue than your lawmaker, but if you don't know an answer to a question, just say so. Be sure to get back to the legislator as soon as possible with the right answer or data.

Be a good listener. Give your legislator the chance to ask questions. Your legislator's views may differ from your own, but at least let him or her know you are willing to hear them.

Put a local angle on the issue. Let the legislator know how a specific issue affects you, your city, town, state, or region.

Press for a commitment. If the legislator is not in opposition to your views, press him or her for commitment. Request support for the specific thing you want.

Bring a fact sheet stating your position. Hand it to the legislator *at the end* of your meeting. Be sure to include a name and phone number on the sheet so the legislator or his or her aides will have a contact for more information.

Write a follow-up letter. After your visit, write a letter thanking your legislator for his or her support for your issue or for any other efforts on your behalf that he or she has done. This will also give you a chance to summarize the key points you made in the meeting and include any information that may have been requested.

Don't forget the staff. If you cannot see your legislator, meet with one of his or her staff. Remember that these aides advise and make recommendations to their bosses about what to do on legislation.

By following these guidelines, your next letter, phone call, or meeting with your members of Congress could help make our environment a little better for everyone.

DISPOSAL OF HOUSEHOLD HAZARDOUS WASTE
Prepared by Linda Wagenet, Ann Lemley, and Judy Stewart

Many products found in your kitchen, bathroom, garage, or garden shed are potentially hazardous substances. Because of their chemical nature, they can poison, corrode, explode, or burst into flame when handled improperly. When discarded, they are household hazardous wastes.

Motor oil, paints, pesticides, antifreeze, wood preservatives, batteries, and some household cleaners contain solvents, petroleum products, heavy metals, or other toxic chemicals. When these products are dumped in the trash or poured down the drain, their hazardous chemicals can injure other people or contaminate drinking water sources.

Motor oil. By law, an auto service or repair station, or any station that changes oil, must accept used motor oil for recycling. To dispose of used motor oil, which contains hazardous petroleum products, first drain the oil through a funnel into a clean container that can be tightly sealed. Then deliver it to your local recycling center. Your Cooperative Extension association or local health department can help you find the nearest center.

Pesticides. Homeowners often store unused pesticides for many years. Do not use the following banned pesticides: aldrin, chlordane, DBCP, DDT, deilfrin, endrin, heptachlor, kepone, indane, mirex, slivex, toxaphene, 2,4,5-T. Check with your local Cooperative Extension association, Department of Environmental Conservation, or Department of Health for names of other pesticides banned by your state.

Some pesticides have been banned by the government because they are a health risk. Waste pesticides that have been banned, as well as unopened containers of household pesticides, usually can be returned either to the store from which they were purchased or directly to the manufacturer.

Special precautions should be taken when disposing of empty pesticide containers. Rinse the container three times with water and use the rinse water in the same manner the original pesticide was intended. Then wrap the container securely in plastic or newspaper and dispose of it with other household refuse.

Aerosols. Only discard completely empty aerosol cans. Before discarding, spray the contents outside and away from humans or animals

---

From Cornell Cooperative Extension, New York State College of Human Ecology, 1989.
NOTE: *The information given here has been carefully documented and reviewed. The suggestions advise safe disposal of products using the best technology available at the present time. The authors assume no liability for the effectiveness or results of the procedures described. Recycling or hazardous-waste collection days are the recommended disposal methods for all household hazardous wastes.*

until the can is completely empty. Then discard the can with household refuse. Since the contents are under pressure, never put aerosol cans in a trash compactor or incinerator, even when empty.

To Prevent Drinking Water Contamination

Remember the following rules for potentially hazardous products.

Never bury household waste.

Never dump waste along the side of the road.

Never pour waste into a street drain or storm sewer.

Use only the amount of product that is absolutely necessary.

Use household hazardous products as infrequently as possible.

Advice for Disposing of Potentially Hazardous Waste

The best advice is, *don't*. If you cannot use up the product, think of someone who can. A neighbor, school, youth group, church, or service organization may be very pleased to share your leftover cleaning solution, paint, fertilizer, gasoline, or insect spray. When shopping, buy only the house, yard, and automobile products you need, and purchase these products in quantities that you will use up. Finally, consider less toxic alternatives for products that are potentially hazardous. If you must dispose of household product, consider the following guidelines.

Use collection days for household hazardous waste whenever possible.

Do not mix wastes: this could create a violent reaction or form a more hazardous product.

Do not put liquid waste in the trash.

Do not discard large quantities of household waste at one time—for example, a shelf of old paints or several gallons of used antifreeze.

Remember: Recycle a product or hold it for a collection day whenever possible!

Disposal Instructions

The table on page 207 offers general disposal recommendations for potentially hazardous products likely to be in your home. The instructions that follow are a key to the methods recommended in the table. When more than one disposal method is given, the first is preferred and the others should be used when the first is not feasible.

Household refuse: Discard with household refuse that is carried to a sanitary landfill or municipal incinerator. Rinse containers with lots of water before discarding in trash.

Wrapped refuse: Wrap in newspaper and then plastic before adding to household refuse.

Small amount down drain: Pour small quantities down drain (or toilet) with plenty of water; rinse container thoroughly, then discard as household refuse. If the capacity of the sewage or septic system is small or if large amounts are to be disposed of, recycle or wait for hazardous waste collection day.

Solidify/double wrap: Solidify with absorbent material such as kitty litter, sawdust, charcoal, or sand. Allow to dry, then double wrap in plastic and discard with household refuse.

Evaporate/double wrap: Allow to evaporate outside, away from children or pets; double wrap container in plastic, then discard with household refuse.

DISPOSAL OF HOUSEHOLD HAZARDOUS WASTE

Special recycling center: Take product to a special recycling facility, or return to manufacturer.

Do not discard—collection day: Do not discard; store for community household hazardous waste collection or call the local Cooperative Extension association, Department of Environmental Conservation, or Department of Health for specific disposal instructions.

Item	Disposal Instructions
Personal Care	
Cosmetics	Household refuse
Hair permanent solution	Small amount down drain
Hair straightener	Small amount down drain
Medicines, liquid	Small amount down drain
Medicines, nonliquid	Small amount down drain; wrapped refuse
Nail polish	Solidify/double wrap; evaporate/double wrap
Nail polish remover	Do not discard—collection day; evaporate/double wrap; small amount down drain
Perfume	Small amount down drain
Shaving lotion	Small amount down drain
Shoe polish	Household refuse
Shoe dye	Do not discard—collection day
Home Care and Maintenance	
Batteries (D cell or smaller)	
Mercury	Special recycling center; wrapped refuse
Hearing aid	Special recycling center (hospital or hearing aid center)
Cleaners (Do not mix chlorine and ammonia-base cleaners)	
Ammonia base	Small amount down drain
Basin, tub, and tile	Small amount down drain
Bleach	Small amount down drain
Drain (lye base)	Small amount down drain

(continued)

Item	Disposal Instructions
Home Care and Maintenance—continued	
Cleaners—*continued*	
Powder/abrasive	Household refuse
Mildew (fungicide)	Do not discard—collection day; wrapped refuse
Oven (lye base)	Small amount down drain
Toilet bowl	Small amount down drain
Upholstery/rug (detergent base)	Small amount down drain
Upholstery/rug (solvent base)	Do not discard—collection day; evaporate/double wrap; small amount down drain
Window	Small amount down drain
Disinfectant	Small amount down drain
Dry-cleaning fluid	Do not discard—collection day; evaporate/double wrap
Fiberglass (epoxy resin)	Do not discard—collection day; solidify resin and hardener, then treat as wrapped refuse
Fluorescent lamp ballast (manufactured prior to 1978 or without label stating it contains no PCBs)	Do not discard—collection day
Glue (solvent base)	Do not discard—collection day; evaporate/double wrap
Mothballs	Do not discard—collection day
Paints or stains	
Latex	Evaporate/double wrap; solidify/double wrap
Oil	Do not discard—collection day
Primer	Do not discard—collection day
Rust	Do not discard—collection day
Stain	Do not discard—collection day
Varnish	Do not discard—collection day
Wood preservative	Do not discard—collection day
Paint remover	Do not discard—collection day; evaporate/double wrap

DISPOSAL OF HOUSEHOLD HAZARDOUS WASTE

Item	Disposal Instructions
Paint thinner	Do not discard—collection day; evaporate/double wrap
Paintbrush cleaner (phosphate base)	Small amount down drain
Paintbrush cleaner (solvent base)	Do not discard—collection day
Paint/varnish stripper (lye base)	Small amount down drain
Polishes	
Copper	Evaporate/double wrap; solidify/double wrap
Floor	Do not discard—collection day
Furniture (solvent base)	Do not discard—collection day
Silver	Solidify/double wrap; evaporate/double wrap
Rust remover (phosphoric acid base)	Small amount down drain
Smoke detector (ionization type)	Special recycling center; return to manufacturer
Spot remover (solvent base)	Do not discard—collection day; evaporate/double wrap
Turpentine	Do not discard—collection day; evaporate/double wrap (store sludge for collection day)

Automobile and Motor Care

Item	Disposal Instructions
Antifreeze (less than 1 gal.)	Small amount down drain; special recycling center
Automatic transmission fluid	Special recycling center
Batteries	Special recycling center
Brake fluid	Special recycling center
Carburetor cleaner	Do not discard—collection day
Degreasing chemicals	Do not discard—collection day
Diesel fuel	Special recycling center
Enamel	Do not discard—collection day; evaporate/double wrap
Fuel oil	Special recycling center
Gasoline	Do not discard—collection day; evaporate/double wrap

(continued)

Item	Disposal Instructions
Automobile and Motor Care—continued	
Kerosene	Special recycling center
Light lubricating oil	Special recycling center
Motor oil	Special recycling center
Polishes or waxes	
Automobile	Do not discard—collection day; evaporate/double wrap
Chrome (solvent base)	Do not discard—collection day; evaporate/double wrap
Windshield washer fluid	Small amount down drain
Lawn and Garden Care	
Fertilizer, liquid (less than 1 gal.)	Wrapped refuse
Fertilizer, dry (less than 5 lb.; less than 25 lb. lawn fertilizer/pesticide)	Wrapped refuse
Mothballs	Do not discard—collection day
Pesticides	
Fungicides	Do not discard—collection day
Swimming pool chemicals (undiluted)	Do not discard—collection day
Poison, rat/mouse (arsenic base)	Do not discard—collection day
Poison, rat/mouse (warfarin base)	Wrapped refuse
Roach/ant killer	Do not discard—collection day
Weed killers	Do not discard—collection day
Garden insecticides	Do not discard—collection day

DIRECTORY OF ENVIRONMENTAL GROUPS

Citizens Clearinghouse for Hazardous Waste
P.O. Box 926
Arlington, VA 22216
(703) 276-7070

Provides assistance to citizens and grass roots groups fighting hazardous waste problems. Communicates information through a variety of environmental publications; offers workshops to help communities organize and solve environmental problems; provides individuals, groups, and communities with scientific and technical assistance in interpreting and using technical information.

The Cousteau Society
930 W. 21st St.
Norfolk, VA 23517
(804) 627-1144

An international, nonprofit, membership-supported organization dedicated to the protection and improvement of the quality of life. Current focus: Informing and alerting the public through television/film, research, lectures, books, and publications focusing on humanity's interaction with the environment. Members: 300,000. Volunteer programs: Many volunteers work in the Norfolk office.

Defenders of Wildlife
1244 19th Street NW
Washington, DC 20036
(202) 659-9510

A national, nonprofit organization whose goal is to preserve, enhance, and protect wildlife and preserve the integrity of natural ecosystems. Current focus: Protecting and restoring habitats and wildlife and promoting wildlife appreciation and education. Members: 80,000-plus members and supporters. Volunteer programs: An activist network consisting of more than 6,000 individuals. Contact Cindy Shogan at the above address for more information.

Ducks Unlimited
One Waterfowl Way
Long Grove, IL 60047
(312) 438-4300

Purpose: To raise money for developing, preserving, restoring, and maintaining the waterfowl habitat on the North American continent and to educate the general public concerning wetlands and waterfowl management. Current focus: Providing leadership and helping fund the North American Waterfowl Management Plan. Members: 550,000. Volunteer

Parts of this listing are from the 1989 Conservation Directory *of the National Wildlife Federation. Reprinted by permission.*

programs: Has more than 5,000 volunteer committees in all parts of the U.S., Canada, and Mexico. Contact the above address for regional information.

Earth Island Institute
300 Broadway, Suite 28
San Francisco, CA 94133
(415) 788-3666

Purpose: To initiate and support internationally oriented action projects for the protection and restoration of the environment. Current focus: Stopping the killing of dolphins, rainforest protection, and fostering environmental protection and restoration in Central America. Members: 25,000. Members receive the quarterly international newsmagazine, *Earth Island Journal*. Volunteer programs: Program internships are available in each project area, as are journalism internships.

Environmental Action Foundation
1525 New Hampshire Ave. NW
Washington, DC 20036
(202) 745-4870

Does technical/legal research and organizing around several solid waste issues, including source reduction of solid waste, citizen suits and participation, corrective action and closure, underground storage tanks, and connections between Resource, Conservation, and Recovery Act and Right-to-Know.

Environmental Defense Fund Inc.
257 Park South
New York, NY 10010
(212) 505-2100

Committed to a multidisciplinary approach to environmental problems, combining the efforts of scientists, economists, and attorneys to devise practical, economically sustainable solutions to these problems. Current focus: Toxic chemicals and waste, water resources and land use, energy and air quality, global atmospheric issues, wildlife and habitats. Members: 100,000. Volunteer programs under development.

Environmental Policy Institute
218 D St. SE
Washington, DC 20003
(202) 544-2600

An independent, nonprofit public interest advocacy organization, founded in 1972, that is dedicated to influencing national and international public policies affecting natural resources through research, public education, lobbying, litigation, and organization of coalitions that are economically, politically, and geographically diverse. Helps citizens influence policies by informing them how their views can be most effectively voiced in Washington. Primary areas of focus: Energy, water, agriculture, toxics, national security as affected by natural resources, and public health. Merged with the Environmental Policy Center in 1982.

Friends of the Earth
218 D St. SE
Washington, DC 20003
(202) 544-2600

Dedicated to the conservation, protection, and rational use of the earth. Engages in lobbying, litigation, and public information on a variety of environmental issues. Publishes the magazine *Not Man Apart*. Affiliated with 33 FOE groups around the world and recently merged with the Oceanic Society and the Environmental Policy Institute. Current focus: Ozone depletion, river protection, tropical deforestation. Members: 12,000. Volunteer programs: Internships available.

DIRECTORY OF ENVIRONMENTAL GROUPS

Greenpeace, USA Inc.
1436 U St. NW
Washington, DC 20009
(202) 462-1177

An international organization dedicated to protecting the natural environment. Works to shield the environment from nuclear and toxic pollution, halt the slaughter of whales and seals, stop nuclear weapons testing, and curtail the arms race at sea. Also campaigns against the mining and reprocessing of nuclear fuel, the exploitation of Antarctica, and the destruction of marine resources through indiscriminant fishing. Members: three million supporters worldwide, offices in 22 countries.

Izaak Walton League of America
1401 Wilson Blvd., Level B
Arlington, VA 22209
(703) 528-1818

A national conservation organization started in 1922 that works toward the protection of America's land, water, and air resources. Current focus: Acid rain, clean water, stream protection, Chesapeake Bay cleanup, outdoor ethics, public land management, soil erosion, clean air, and waterfowl/wildlife protection. Members: 50,000 (400 chapters). Volunteer programs: Save Our Streams, Wetlands Watch, Oregon Public Lands Restoration Task Force, chapter programs.

National Audubon Society
950 3rd Ave.
New York, NY 10022
(212) 832-3200

Purpose: To conserve plants and animals and their habitats; promote rational strategies for energy development and use, stressing conservation and renewable resources; protect life from pollution, radiation, and toxic substances; further the wise use of land and water; seek solutions for global problems involving the interaction of population, resources, and the environment. Members: 550,000. Volunteer programs: Annual Christmas Bird Count and Breeding Bird Census; Audubon Activist Network; internships; Citizen's Acid Rain Monitoring Network; involvement through local chapters, state and regional offices.

National Coalition against the Misuse of Pesticides
530 7th St. SE
Washington, DC 20003
(202) 543-5450

A nonprofit membership organization formed in 1981 to serve as a national network committed to pesticide safety and the adoption of alternative pest-management strategies that reduce or eliminate a dependency on toxic chemicals. Provides useful information on pesticides and alternatives to their use; publishes newsletters "Pesticides and You" and "NCAMP's Technical Report."

National Wildlife Federation
1400 16th St. NW
Washington, DC 20036-2266
(202)797-6800

Promotes the wise use of natural resources and protection of the global environment. Distributes periodicals and educational materials, sponsors outdoor education programs in conservation, and litigates environmental disputes in an effort to conserve natural resources and wildlife. Current focus: Forests, energy, toxic pollution, environmental quality, biotechnical fisheries and wildlife, wetlands, water resources, public lands. Members: More than five million. Volunteer programs: The National Wildlife Federation oversees 51 state affiliates, all of which rely on volunteer support.

Natural Resources Defense Council
40 W. 20th St.
New York, NY 10011
(212) 727-2700

Purpose: To protect natural resources and improve the quality of the human environment. Combines legal action, scientific research, and citizen education in a highly effective environmental protection program. Major accomplishments have been in the area of energy policy and nuclear safety, air and water pollution, urban environmental issues, toxic substances, and natural resources conservation. Members: 95,000.

The Nature Conservancy
1815 N. Lynn St.
Arlington, VA 22209
(703) 841-5300

An international, nonprofit membership organization committed to preserving biological diversity by protecting natural lands and the life they harbor. Cooperates with educational institutions, public and private conservation agencies. Works with states through "natural heritage programs" to identify ecologically significant natural areas. Manages a system of over 1,000 nature sanctuaries nationwide. Works with Central and South American governments to identify and protect natural lands and tropical rainforests.

The Oceanic Society
218 D St. SE
Washington, DC 20036
(202) 544-2600

Dedicated solely to protecting the oceans for people and wildlife. Through education, research, conservation, and advocacy, promotes the understanding and stewardship of the marine and coastal environment. In 1989 it merged with Friends of the Earth and the Environmental Policy Institute. Current focus: Ocean disposal of wastes. Members: 40,000. Volunteer programs: None in the national office, but volunteers are welcome at the Long Island Sound, San Francisco Bay, and L.A. offices, and the San Francisco-based Expeditions program.

Rainforest Action Network
301 Broadway, Suite A
San Francisco, CA 94133
(415) 398-4404

A nonprofit, activist organization working to save the world's rainforests. Begun in 1984, it works internationally in cooperation with other environmental and human rights organizations on campaigns to protect rainforests. Provides financial support for groups in tropical countries to preserve forest lands. Methods include direct action, grassroots organizing, and media outreach. Current focus: Tropical timber, indigenous peoples, and multilateral development banks. Members: 20,000. Volunteer programs: Internships are available.

Renew America
1400 16th St. NW
Suite 710
Washington, DC 20036

A nonprofit, tax-exempt organization created to move America toward a sustainable society. It is a continuation of Solar Action, which was formed in 1978 to promote the celebration of "Sun Day." It addresses the need for increased natural resource efficiency, including renewable energy, sustainable agricultural practices, water conservation, and the recycling of refined materials. Also designs state and local action programs and conducts legislative oversight on Capitol Hill through the Renew America Scorecard. Members: More than 7,000.

Sierra Club
730 Polk St.
San Francisco, CA 94109
(415) 776-2211

Purpose: To promote conservation of the natural environment by influencing public policy; to practice and promote the responsible use of the earth's ecosystems and resources; to educate and enlist humanity to protect and restore the quality of the natural and human environment. Current focus: Clean Air Act reauthorization, Arctic National Wildlife refuge protection, Bureau of Land Management wilderness/desert national parks protection, toxic waste regulations, global warming/greenhouse effect. Members: 500,000 (57 chapters). Volunteer programs: Extensive opportunities available throughout the country in conservation campaigns and outdoor activities.

Trout Unlimited
501 Church St. NE
Vienna, VA 22180
(703) 281-1100

A nonprofit organization dedicated to the protection of clean waters and the enhancement of trout, salmon, and steelhead fishing. The national office works with Congress and federal agencies for the protection and wise management of America's fishing waters, interacts with other national conservation organizations, sponsors seminars, funds research projects, and administers the nationwide network of chapters. Current focus: Water pollution, environmental abuse, waste management, wild rivers, fish habitat, education. Members: 55,000 (480 chapters). Volunteer programs: All chapters welcome the assistance of volunteers.

Union of Concerned Scientists
26 Church St.
Cambridge, MA 02238
(617) 547-5552

A nonprofit organization founded in 1969 that conducts policy and technical research, public education, and legislative advocacy on issues concerning advanced technologies. A coalition of scientists, engineers, and other professionals concerned with health, safety, environmental, and national security problems posed by the country's nuclear programs. Sponsorship: Over 100,000.

U.S. Public Interest Research Group
215 Pennsylvania Ave. SE
Washington, DC 20003
(202) 546-9707

A nonprofit, nonpartisan environmental and consumer advocacy organization, founded in 1983, which represents the public's interest in the areas of consumer and environmental protection, energy policy, and governmental and corporate reform. Current focus: New legislation for clean air, pesticide and product safety, reducing the use of toxics. Members: 500,000.

Whale Center
3933 Piedmont Ave., Suite 2
Oakland, CA 94611
(415) 654-6621

A nonprofit conservation and education organization working to save whales and their ocean habitat. Oversees conservation projects, including opposing commercial whaling, preventing the drowning of marine mammals in nets, and establishing new national marine sanctuaries; education through the WhaleBus school program and annual whale-watching cruises along the California coast. Publishes the quarterly *Whale Center Journal* and maintains the Whale Center Library of Marine Mammals. Members: 1,300. Volunteer programs: Volunteers are invited to help on conservation projects and in office clerical positions.

The Wilderness Society
1400 I St. NW, 10th Fl.
Washington, DC 20005
(202) 842-3400

A nonprofit membership organization, founded in 1935, devoted to preserving wilderness and wildlife; protecting America's prime forests, parks, rivers, and shorelands; and fostering an American land ethic. Members: 220,000. Welcomes membership inquiries, contributions, and bequests.

World Resources Institute
1709 New York Ave. NW, Suite 706
Washington, DC 20006
(202)638-6300

A policy research center that helps governments, the private sector, and environmental and development organizations address the fundamental question of meeting human needs and nurturing economic growth while preserving natural resources and environmental integrity. Current focus: Tropical forests, biological diversity, sustainable agriculture, energy, climate change, atmospheric pollution, and economic incentives for sustainable development.

World Wildlife Fund
1250 24th St. NW
Washington, DC 20037
(202) 293-4800

A private U.S. organization working worldwide to protect endangered wildlife and wildlands. Its top priority is conservation of the tropical forests of Latin America, Asia, and Africa. Current focus: a campaign to save the African elephant, sustainable development programs in Asia and Africa, the panda management plan with China TRAFFIC—a program that monitors trade in wild plants and animals—and a jaguar preserve in Belize. Affiliated with the international WWF network, which includes national organizations and associates in 26 countries across five continents. Membership: 600,000 plus.

Zero Population Growth
1400 16th St. NW, Suite 320
Washington, DC 20036
(202) 332-2200

A nonprofit membership organization that works to achieve a sustainable balance between the earth's population, its environment, and its resources. Primary activities include publishing newsletters and research reports, developing in-school population education programs, and coordinating local and national citizen action efforts. Members: 20,000. Volunteer programs: Legislative Alert Network (congressional support for critical population/environmental legislation); Media Targets (volunteers write to media to encourage better reporting); Teacher's Pet Project (volunteers help with teacher-training workshops).

INDEX

Page references in *italics* indicate photographs or tables.

A
ACEEE. *See* American Council for an Energy Efficient Economy
Acid moss, 5–6
Acid rain
 congress and, 9
 Earth Care actions against, 10–12
 forest devastation and, 4–5
 legislation on, 9
 limestone and, 10–12
 sources of, 3
 threat of, 3
 worsening effects of, 6–7
Acid soils, 5–6
Adirondack Park, New York, acid rain and, 6–7
Aerosols, disposal of, 205–6
 instructions for, 207–10
Africa
 cleanup ritual in Nigeria, 120–22, *121*
 forest replenishment in, 151–52
Agriculture, sustainable, xiii
Air-conditioning, automobile
 chlorofluorocarbons and, 47
 system redesign for, 49
Air pollutants, 15
Air pollution. *See also specific pollutants*
 in Chesapeake Bay, 7–8
 as world problem, 13–16
 in world's big cities, 13
Alar, 60
Aluminum, acid rain and, 5–6
American Council for an Energy Efficient Economy (ACEEE),
 energy efficiency and, 26–27
America's Smog Crisis, 14
AT&T, paper-recycling program of, 115, 116
Automobile emissions, air pollution and, 21
Automobiles. *See also* Solar cars
 air-conditioning in
 chlorofluorocarbons and, 47
 system redesign for, 49
 energy efficiency and, 31
 reduction of air pollution and, 28
Aylesworth, Art, bluebirds and, 181–82, *181*

B
Baking soda, gaseous emissions and, 6
Batteries
 recycling of, 137–38
 toxic waste and, 136–37
 community programs on, 138
Benirschke, Rolf, San Diego Zoo and, 178–80, *179*
Benzene, 73
Berkeley, California, recycling dropoff program in, 107–8
Bicycles, 21–22
BioAct EC7 as chlorofluorocarbon substitute, 49–50
Biodegradable diapers, 94
Biodegradable plastics, 82–83
Biodynamic, definition of, 62
Birds. *See also* Bluebirds
 deforestation and, 155
Bluebirds, 181–82, *181*
Blue Box program, 118–19
Brazil, deforestation in, 144
Buildings, smart, energy efficiency and, 26
BUND. *See* German Association for Environmental and Nature Protection
Bureau of Land Management, 191
Bureau of Reclamation, 191
Buxman, Paul, pesticides and, 69, *69*

C
California
 curbside pickup recycling program in, 108–9
 recycling dropoff program in, 107–8
 solid waste legislation in, 101
 wildlife protection and, 160–64
Californians for Parks and Wildlife (CALPAW), formation of, 161
CALPAW. *See* Californians for Parks and Wildlife
Cancer, pesticides and, 73, 74
Carbon dioxide, greenhouse effect and, 24, 26–27
Carbon monoxide, 15–16
 in Paris, 13
Cars. *See* Automobiles; Solar cars
Center for Plastics Recycling Research, 87
CFC. *See* Chlorofluorocarbons
Chemicals. *See also* Pesticides
 chlorofluorocarbon alternatives, 48–49

Chemicals (*continued*)
  industry and pollution with, ix–x
  ozone-depleting, 46–47
ChemLawn, pesticides used by, 73–74
Chesapeake Bay, pollution in, 7–8
Cheyenne Bottoms Wildlife Area, 182–84, *183*
Chlorofluorocarbons (CFCs)
  air pollution and, 32
  alternatives to, 45–46
    chemicals as, 48–49
  atmospheric ozone and, 44–49
  in cooling systems, 47–48
  corporations and, 45
  greenhouse effect and, 25
  phase out of, 51–52
  reduction of air pollution and, 28
  refrigerant loss and, 47–48
Chlorofluorocarbon treaty, under Montreal Protocol, 37
Chloroform, 73
Chlorothalonil, 74
  toxic epidermal necrolysis and, 73
Christmas trees, recycling of, 156
Citizens Clearinghouse for Hazardous Waste, 211
Clams, giant, 171, *172*, 173–75
Climate changes, greenhouse effect and, 24
Coal, efficient burning of, 28
Coal industry, acidified waters and, 7
Coast Guard, 192
Cogeneration, energy efficiency and, 28, 29
Computers, solid waste and, 116
Congress
  addresses for, 203
  appropriations of, 200
  committees of, 199
  conference committee of, 200
  floor action of, 200
  hearings of, 199
  influencing, 198–99
  markup of, 199–200
  personal meetings with, 203–4
  process of, 199–200
  telegrams to, 202–3
  telephone calls to, 202
  telephone numbers for, 203
  writing letters to, 199–201
    follow-up, 201–2
    sample, 201
Connecticut, solid waste legislation in, 101–2
Cornstarch, biodegradable plastics and, 82–83
Corporations. *See also* Industry, chemical assault on environment by,
  chlorofluorocarbons and, 45
  recycling by, 115–16
Costa Rica, answer to deforestation, 147–49
The Cousteau Society, 211
CPRR. *See* Center for Plastics Recycling Research

D
Daconil, toxic epidermal necrolysis and, 73
Daminozide, 60
Defenders of Wildlife, 211
Deforestation. *See* Tropical forests, destruction of
Department of Agriculture, 191–92
Department of Commerce, 192
Department of Defense, 192
Department of Energy, 192
Department of the Interior, 191
Department of Transportation, 192
Development, sustainable, xiii
Diapers, 89–94
  biodegradable, 94
  comparison of costs of, *90*
  report on, 90–91
  solutions to, 93
Diaper services, 92–93
  cost of, *90*
Diesel fuel, 22
Diflubenzurone, 72
Disposal instructions for household hazardous waste, 207–10
Double-paned windows, energy efficiency and, 26
Drinking water, prevention of contamination of, 206
Dry forests. *See* Tropical forests
Ducks, plight of, 176
Ducks Unlimited, 211–12
Dust as air pollutant, 13

E
Earth care
  optimism about, xv
  trend toward, xii–xiii
Earth care actions
  acid rain and, 10–12
  atmospheric ozone and, 49–53
  energy efficiency and, 28
  greenhouse effect and, 31–43
  industrial gaseous emissions and, 6
  landfills and, 110–13
  in Nigeria, 120–22, *121*
  ozone layer and, 46
  pesticides and, 58, 61–75
  plastics and, 78, 80, 85–94
  polluted rivers and, 102–5
  recycling programs, 106–10
    in Ontario, 118–19
  smog and, 16–22
  solid waste and, 100–122
    legislation on, 100–102
  toxic waste and, 131–143
  water pollution and, 11
  wilderness and, 157–64
  wildlife and, 171–85
Earth Day
  in 1970, xi–xii
  twentieth anniversary of, ix
Earth Island Institute, 212
Earth's temperature. *See* Greenhouse effect
Ecology Action
  curbside pickup program of, 108–9
  founding of, 106
  recycling dropoff program of, 107–8
EDF. *See* Environmental Defense Fund, Inc.
Electricity, efficient use of, 31–33
Electric power plants, air pollution and, 32
Energy, renewable sources of, 40–41
Energy consumption
  air pollution and, 28
  in United States, 30

INDEX

Energy efficiency
  earth care actions for, 28
  greenhouse effect and, 26–27
  in homes, 28
  nonfossil fuels and, 28–29
  options for, 29
  promotion of, 27
  renewable technologies available, 29
  United States backsliding in, 30
Eno River Valley, 157–60, *159*
Environment, chemical assault on, ix–x
Environmental Action Foundation, 212
Environmental Defense Fund, Inc., 212
Environmental groups, directory of, 211–16
Environmental Policy Institute, 212
Environmental problems
  action alert for, 189–93
  calling attention to
    through newspapers, 194–96
    through radio and television, 196–98
Environmental Protection Agency (EPA)
  description of, 191
  toxic waste and, 124–25
EPA. *See* Environmental Protection Agency
Europe, plastic packaging in, 83

F

FAO. *See* Food and Agriculture Organization
Farming. *See also* Organic farming; Sustainable agriculture
  deforestation and, 146
  organic methods of, interest in, 61–64
  water pollution and, xiii
FDA. *See* Food and Drug Administration
Federal agencies, 189, 191–92
  daily newspaper for, 189–90
  power of, 189
  writing to, 192–93
Federal Energy Regulatory Commission, 192

Federal Highway Administration, 192
Federal Register, 190
  use of, 190–91
Fenoxycarb, in lieu of pesticides, 72
Fish, plight of, 176
Florida, solid waste legislation in, 101
Fluorescent bulbs, energy efficiency and, 26–27
Food and Agriculture Organization (FAO), pesticides and, 65
Food and Drug Administration (FDA), pesticides and, 59
Forests
  designated as wilderness, 154
  disappearance of, 155
  export of felled trees, 154–55
  fate of, 153
Freiburg, West Germany
  Department for Environmental Affairs hotline, 139
  toxic waste and, 138–41
Freons, reduction of air pollution and, 28
Friends of the Earth, 212
Fruits, cancer-causing pesticides in, x
Fuel. *See specific fuels*
Furnaces, energy efficiency and, 35

G

Gamez, Rodrigo, *149*
Garbage. *See* Solid waste
Garton, Jan, wetland habitat and, 182–84, *183*
Gas, natural, reduction of pollution and, 28
Gaseous emissions, baking soda and, 6
Gasoline
  air pollution and, 22
  efficient use of, 31
GEMS. *See* Global Environment Monitoring Systems
German Association for Environmental and Nature Protection (BUND), toxic waste and, 140
Giant clams, 171, *172,* 173–75

Global Environment Monitoring Systems (GEMS), smog and, 14
Global Greenhouse Network, founding of, 37
Global warming. *See* Greenhouse effect
Golf courses, pesticides applied to, 73–75
Grapefruit rinds, atmospheric ozone and, 49–50
Greenhouse effect, 23–24, 26–30
  Congress and, 34–35
  deforestation and, x
  earth care actions and, 31–43
  gases contributing to, 24–27
  government responsibility in, 34
  individual responsibility in, 35–36
  international responsibility in, 36–37
  Los Angeles and, 42–43
  motor vehicle emissions and, 27–28
  solution to, 40–41
  state responsibility in, 36
Greenpeace, 213

H

Habitat destruction, wildlife endangerment by, x
Hazardous waste, household, disposal of, 205–10. *See also* Toxic waste
High-temperature, gas-cooled reactor (HTGR), greenhouse effect and, 40
Highway deaths, reduction of, in Nigeria, 121–22
Homes, energy efficient, 28
Hormones, synthetic, in lieu of pesticides, 72
Household hazardous waste, 205–6
  disposal of
    advice for, 206
    instructions for, 206–10
HTGR. *See* High-temperature, gas-cooled reactor
Humphrey, Cliff, recycling and, 106–10, *107*

Hydrogen fuel, reduction of air pollution and, 28

I
Illinois, solid waste legislation in, 102
Incinerators for solid waste, 97–99
  in Massachusetts, 101
Industry, chemical assault on environment by, ix–x. *See also specific industries;* Corporations
Insulation. *See* Superinsulation
Internal resource, description of, xiv
Iowa, solid waste legislation in, 102
Izaak Walton League of America, 213

J
Jackson Elementary School students, *131, 134*
  toxic waste cleanup crusade by, 131–36
Japan, "research whaling" by, 167–70

K
Kissida, John, use of closed landfills and, 110–13

L
Landfills
  alternatives to, 111
  closed, use of, 110–13
  ground-water contamination from, 112
  leachate from, 112
  methane gas from, 111–12
  solid waste and, 97
Lawn-care companies, alternatives offered by, 74
Lawns, pesticides applied to, 73–75
Leachate from landfills, 112
Lead, air pollution and, 15–16
Legislation
  on plastics recycling, 80–81
  on solid waste, 100–102
    in California, 101
    in Connecticut, 101–2
    in Florida, 101
    in Illinois, 102
    in Iowa, 102
    in Pennsylvania, 101
    in Rhode Island, 102
    in Washington, 102
Lighting, energy efficiency and, 26–27
Limestone treatment of polluted waters, 10–12
Litter. *See* Solid waste
Loetscher, Ila, ridleys and, 180–81, *180*
Logging. *See* Timber industry
Los Angeles, tree planting program in, 42–43
Lumber, plastic, 87–88

M
Mancozeb, 74
Massachusetts
  acid rain and, 7
  Solid Waste Act of, 101
Mass transit, energy efficiency and, 27–28
Methane gas
  greenhouse effect and, 25
  from landfills, 111–12
Minnesota, plastics ordinance in, 85–86
Mobile air-conditioning. *See* Air-conditioning, automobiles
Modesto, California, recycling dropoff program in, 108–9
Montreal Protocol, 44
  phase out of chlorofluorocarbons and, 51
Moss, acid, 5–6
Mothers and Others for Pesticide Limits, 57–58
  starting your own chapter of, 58
  symposium held by, 60–61
Motor oil, disposal of, 205, 209–10
Motor vehicle emissions, global warming and, 27–28
Mount Mitchell, North Carolina, forest devastation in, 4–5

N
NAPAP. *See* National Acid Precipitation Assessment Program
National Acid Precipitation Assessment Program (NAPAP), acid rain and, 8–9
National Audubon Society, 213
National Coalition against the Misuse of Pesticides, 213
National Environmental Sanitation Exercise, of Lagos, Nigeria, 120–22, *121*
National Marine Fisheries Service, 192
National Oceanic and Atmospheric Administration, 192
National Park Service, 191
National Wildlife Federation (NWF), 213
  environment and, x
Nations, greenhouse effect and, 36–37
Natural gas, reduction of pollution and, 28
Natural Resources Defense Council (NRDC), 214
  report on pesticides by, 56
The Nature Conservancy, 214
New Jersey
  mandated recycling in, 82
  pay-by-container trash-collection system in, 113–15
News media, calling attention to environmental problems and, 194–98
Newspapers. *See* News media
New York State, acid rain and, 6–7
Nigeria
  cleanup ritual in, 120–22, *121*
  reducing highway deaths in, 121–22
  toxic waste in, 142
Nitrogen, importance of, xiii
Nitrogen dioxide, as air pollutant, 16
Nitrogen oxides
  acid rain and, 7–8
  greenhouse effect and, 25
NOAA. *See* National Oceanic and Atmospheric Administration
Nonfossil fuels, energy efficiency and, 28–29

INDEX

North Carolina
  Eno River Valley in, 157–60, *159*
  forest devastation in, 4–5
No-spray, definition of, 62
NO_x agreement, 37
NRDC. *See* Natural Resources Defense Council
Nuclear reactors
  greenhouse effect and, 37–39
  new designs in, 39–40
NWF. *See* National Wildlife Federation
Nygard, Margaret, Eno River Valley and, 157–60, *159*

O

The Oceanic Society, 214
Oceans, plastic trash and, 77–78
Office of Conservation and Renewable Energy, 192
Office of Surface Mining, 191
Ontario, recycling program of, 118–19
Orange rinds, atmospheric ozone and, 49–50
Oregon, polluted rivers in, 102–5
ORFA factory, for solid waste, 95
Organic, definition of, 62
Organic farming
  false claims of, 63–64
  increase in, 61–64
  myth versus reality of, 70
Ozone
  acid rain and levels of, 5, 7
  greenhouse effect and, 25
Ozone, atmospheric
  chlorofluorocarbons and, 44–49
  earth care actions for, 46, 49–53
  orange and grapefruit rinds and, 49–50
  Vermont and, 52–53
Ozone-depleting chemicals, stocks of, 46–47
Ozone layer. *See* Ozone, atmospheric
Ozone smog, air pollution and, 32

P

Paper, recycling programs for, 115–16
Parks, pesticides applied to, 73–75
Pay-by-container trash-collection system, 113–15
PCL. *See* Planning and Conservation League
Pennsylvania
  acid rain and, 7
  solid waste legislation in, 101
Pest control, strategies for, 64–66
Pesticides
  alternatives offered by lawn-care companies, 74
  cancer-causing, in fruits and vegetables, x
  cutting back on, 66–67
    commitment to, 70–71
    in Indonesia, 64–66
    perceived obstacles in, 67
    realities in, 67–68
    reliable information in, 68–70
    surprises in, 68
  disposal of, 205, 210
  earth care actions and, 58, 61–75
  Food and Drug Administration and, 59
  inert ingredients in, 73
  on lawns, 73–75
  mothers against, 54–61
  Natural Resources Defense Council report on, 56
  safety claims by manufacturers of, 74–75
  symposium held on, 60–61
  synthetic hormones in lieu of, 72
PET. *See* Polyethylene teraphthalate
Planning and Conservation League, Proposition 70 and, 161
Plastic(s)
  diapers, 89–94
  earth care actions and, 80
  European packaging with, 83
  pollution by, 76–84
    earth care actions and, 78
    marine environments and, 77
    solutions to, 78
    United Nations and, 77
    wildlife deaths and, 78–79
  polyethylene, 81
  polyethylene teraphthalate, 81
  recycled, 79
  recycling of, 80–83
    into lumber, 87–88
    starch-based biodegradable, 82
    objections to, 82–83
    throwaway, ordinance against, 85–86
  view on
    future, 84
    immediate, 83–84
Plastics industry, recycling by, 80, 81
Pollution. *See also* Air pollution; Water pollution
  controls, industrial, 28
  environmental, by industry, ix–x
  mountaintops and, 5
  plastics and, 76–84
Polyethylene plastics, recycling of, 81
Polyethylene teraphthalate (PET), bottles made of, 81
Polymer Products, recycling by, 81–82
Proposition 70
  Californians for Parks and Wildlife and, 161
  description of, 161–63
  implementation of, 163
  opposition to, 163
  wildlife protection and, 160–64, *162*

R

Radio. *See* News media
Rainforest Action Network, 214
Rain forests. *See* Tropical forests
Rapeseed oil as tractor fuel, 18
RCRA. *See* Resource Conservation and Recovery Act
Recycling
  as alternative to landfills, 111
  of batteries, 137–38
  of Christmas trees, 156
  Cliff Humphrey and, 106–10, *107*
  curbside pickup program for, 108–9
  dropoff program for, 107–8
  energy conservation and, 32–33
  in New Jersey, 82
  Ontario program for, 118–19

Recycling (*continued*)
  of paper, 115–16
  of plastics, 80–83
    legislation on, 80–81
    into lumber, 87–88
    programs for, 106–10
  of solid waste, 99–100
  of tires, 117–18
Redford, Robert, 54–55
Refrigerant loss, chlorofluorocarbons and, 47–48
Regeneration, description of, xiii
Renew America, 214–15
Resource Conservation and Recovery Act, toxic waste and, 124–25
Rhode Island, solid waste legislation in, 102
Rice, pesticides and, 64–66
Ridleys, 180–81, *180*
Rivers, pollution of, 102–5
Rogue River, pollution of, 102–5

S
Safe Drinking Water Act, violations of, ix
San Diego Zoo, 178–80, *179*
Save-the-Whale movement, 165–66
Seabirds, 177–78, *178*
Sea turtles
  endangered, 175
  ridleys, 180–81, *180*
Sierra Club, 215
Slayden, Fred, Virgin Island wildlife and, 184–85
Smart buildings, energy efficiency and, 26
Smog
  crisis
    in America 1988, 14
    in London 1952, 13
  Los Angeles plan for banishment of, 16–17
  Northeastern states plans for reduction of, 17–19
Smoke from coal fires, London smog and, 13
SO_2 agreement, 37
Soda ash, gaseous emissions and, 6

Sodium bicarbonate, gaseous emissions and, 6
Soil, acid, 5–6
Soil Conservation Service, 192
Solar cars, 19–21
Solid waste, 95–97
  archaeologists and, 98
  cleanup of, in Nigeria, 120–22, *121*
  earth care actions and, 100–122
  incinerators and, 97–99, 101
  landfills and, 97, 110–13
  legislation on, 100–102
  paper, 115–16
  pay-by-container collection system, 113–15
  problems in United States, 79–80
  recycling and, 99–100, 101–2, 106–10
    by corporations, 115–16
    in Ontario, 118–19
  river pollution and, 102–5
  tires, 117–18
Speed limits, energy efficiency and, 27, 31
St. Croix, wildlife protection in, 184–85
Starch-based biodegradable plastics, 82–83
State responsibilities, in global warming and, 36
Stopps, Elanor, seabirds and, 177–78, *178*
Streep, Meryl, actions against pesticides by, 54–55, *55*
Sulfur dioxide
  acid rain and, 7
  air pollution and, 13, 15
Superinsulation, greenhouse effect and, 26
Suspended particulate matter. *See* Dust; Smoke
Sustainable, definition of, 62
Sustainable agriculture, xiii
Sustainable development, xiii

T
Television. *See* News media
Temperature, earth's. *See* Greenhouse effect

TEN. *See* Toxic epidermal necrolysis
Thermal storage, energy efficiency and, 26
Timber industry, 154
  pressure from, 155
Timber supply crisis, 155–57
Tires, recycling of, 117–18
Tonga, giant clams of, 171, *172*, 173–75
Toronto Conference, 37
Toxic epidermal necrolysis (TEN), pesticides and, 73
Toxic waste, 123–24. *See also* Household hazardous waste
  Africa tough on, 142–43
  batteries and, 136–38
  control of, 128–30
  disposal of, 124–25
  economic inefficiency and, 127
  end-of-pipe policies on, 125–26
  institutional problems and, 127
  liabilities, 127–28
  prevention of, 128–30
  public-policy choices on, 126–27
  reduction of, 124–25
  regulatory ineffectiveness and, 127
  technological pessimism and, 128
  waste-reduction ethic for, 130
Transitional, definition of, 62
Trash. *See* Solid waste
Trees
  global warming and, 33–34
  Los Angeles planting program, 42–43
Tropical forests, 143–44
  destruction of, 144
    air pollution and, 28
    in Brazil, 29
    by burning, 146
    Costa Rica's answer to, 147–49
    global problem of, 146–47
    global warming caused by, x, 29–30
  hope for, 149
  locations of, 145
  loss of, 145–46
  replenishment of, in Africa, 151–52
  saving wildlife habitat in, 150

INDEX

Trout Unlimited, 215
Turtles, sea, 175

U
UDMH, 60
Ultraviolet radiation, increased, 44
Union of Concerned Scientists, 215
United Nations
  greenhouse effect and, 37
  plastic pollution and, 77
United States
  plight of wildlife in, 176–77
  solid waste problems in, 79–80
United States Fish and Wildlife Service, 191
United States Forest Service, 191–92
U.S. Public Interest Research Group, 215
Utilities, energy efficiency and, 27
UV-B. *See* Ultraviolet radiation

V
Vegetables, cancer-causing pesticides in, x
Vermont, actions to save ozone layer by, 52–53
Virgin Islands, wildlife protection in, 184–85

W
Washington, solid waste legislation in, 102
Waterfowl in North America, 176
Water pollution
  in Chesapeake Bay, 7–8
  Earth Care action and, 11
  farming and, xiii
  limestone and, 10–12
  prevention of, 206
Wetland habitat, 182–84, *183*
Whale Center, 215–16
Whales
  endangered, 166–70
  halting the hunt for, 170
  vital role of, 168
Whaling industry, 166–67
  in Japan, 167–70
WHO. *See* World Health Organization
Wilderness, forests designated as, 154. *See also* Forests
The Wilderness Society, 216
Wildlife. *See also specific wildlife*
  earth care actions and, 171–85
  endangered
    by habitat destruction, x
    whales, 165–70
  meter-saved habitat for, 150
  plastic-related deaths of, 78–79
  plight of, in United States, 176–77
  protection of, Proposition 70 and, 160–64, *162*
  victories for, 176–77
Windows, double-paned, energy efficiency and, 26
World Health Organization (WHO), air pollution and, 13
World Resources Institute, 216
World Wildlife Fund, 216

Z
Zero Population Growth, 216